Lecture Notes in Physics

Edited by H. Araki, Kyoto, J. Ehlers, München, K. Hepp, Zürich
R. Kippenhahn, München, D. Ruelle, Bures-sur-Yvette
H.A. Weidenmüller, Heidelberg, J. Wess, Karlsruhe and J. Zittartz, Köln
Managing Editor: W. Beiglböck

321

Itamar Pitowsky

Quantum Probability – Quantum Logic

Springer-Verlag
Berlin Heidelberg GmbH

Author

Itamar Pitowsky
Department of Philosophy, The Hebrew University Jerusalem
Jerusalem, Israel

ISBN 978-3-662-13735-2   ISBN 978-3-540-46070-1 (eBook)
DOI 10.1007/978-3-540-46070-1

© Springer-Verlag Berlin Heidelberg 1989
Originally published by Springer-Verlag Berlin Heidelberg New York in 1989
Softcover reprint of the hardcover 1st edition 1989

2158/3140-543210

Preface

This monograph is a result of a course on the conceptual foundations of quantum mechanics, given in 1986 and 1988 to physics graduate students, at the Racach Institute of Physics at the Hebrew University, and in 1987 to philosophy graduate students at the University of Western Ontario.

While preparing for the course, I was struck by the immense number and variety of approaches to the problem of interpretation of quantum mechanics. The variability presents itself not just on the level of ideology – which is understandable – but also on the level of notations and formulations. It seems that people do not quite agree on the proper way to present the questions, or even what the questions themselves are. Being perhaps overly optimistic, I set out to look for a unifying principle; a formulation that will make it possible to present the serious alternative approaches, and compare them, on a fair common ground.

As a first attempt I chose to examine the concept "correlation" and for simplicity to deal with correlations of events, rather than random variables in general. The result was an article, Pitowsky (1986), which is essentially a comparison between the quantum concept of correlation and correlations as conceived in traditional probability theory. I was surprised at the richness and unifying power of the subject, with the result that the article has grown into this monograph.

The possible range of values of classical correlations is constrained by linear inequalities which can be presented as facets of polytopes, which I call "classical correlation polytopes." These constraints have been the subject of investigation by probability theorists and statisticians at least since the 1930s, though the context of investigation was far removed from physics. The linear constraints in question include Bell's inequalities, Clauser–Horne inequalities and their generalizations. Chapter 2 of this monograph is devoted to the study of classical correlation polytopes, their description in terms of vertices and facets, and their relations to propositional logic and computer science. The subject has applications which transcend quantum physics and even physics in general: It is closely related to the study of the Ising spin model, neural networks, and computational complexity. Some of these applications are indicated in the text, and others, which transcend the scope of this monograph, are indicated in Pitowsky (1988).

Chapter 3 is a similar analysis of quantum correlations in terms of linear inequalities and polytopes. At its center is a theorem which completely characterizes the possible range of values of quantum correlations. The rest of Chapter 3 is devoted to examples where classical constraints are violated by quantum frequencies, most notably the violation of Bell–type inequalities in the Einstein–Podolsky–Rosen experiment.

Classical correlation polytopes are also closely related to classical propositional logic. An argument by analogy, with respect to quantum correlation polytopes, leads directly to quantum logic. Chapter 4 is devoted to the study of quantum logic. In particular I prove that any realistic conception of quantum logic implies a violation of locality. Chapter 5 is an analysis of the hidden variable approach. It includes a detailed construction of local hidden variable theories, based on an extension of classical probability. This generalizes an earlier work of mine and its extension by S.P. Gudder. In particular I address some objections raised in the physics literature.

On the level of interpretation, the analysis of correlations provides for what I believe to be a fair comparison of four basic approaches: The Copenhagen interpretation, the antirealist view, hidden variable theories and quantum logic. No reference is made to the "many worlds interpretation", and mystical views are mentioned only in passing.

This book is a research monograph, and is not intended as a review or a textbook. Consequently, references are made only to those publications which bear directly on the text. Even so, the scope is quite vast, and I cannot pretend to cover all the relevant material. I thus apologize for any omission which results from my ignorance.

Through the years I have benefited from conversations and correspondence with many colleagues and friends. Many thanks are due to Jeffrey Bub who taught me the basic lessons on the "quantum muddle", to the late Peter Moldauer, to Stanley Gudder, David Mermin, Henry Stapp, Malcolm Forster, Abner Shimony, Arthur Fine, Michael Friedman, Roger Cooke, David Malament, Alan Stairs, David Albert, Yemima Ben-Menachem, Mara Beller, Mark Steiner, Benjamin Weiss, and Mendel Sachs.

Parts of this monograph were written while I was visiting the University of

Western Ontario in the fall of 1986 and 1987. I would like to thank the Department of Philosophy for its hospitality, in particular Ray and Bill Demopolous and Lisa and Michael Daws. I would also like to thank Nancy Weber for preparing this manuscript for publication.

This research was supported by a grant from the Israeli Ministry of Absorption and by the Sidney M. Edelstein Center for the History and Philosophy of Science at the Hebrew University, whose assistance is hereby acknowledged.

Last but not least I am grateful to my wife Liora Lurie and my daughters Noga and Michelle for their love, patience and support.

Table of Contents

1. Introduction

In his little book, *The Character of Physical Law*, Richard Feynman makes the following statement:

There was a time when the newspapers said that only twelve men understood the theory of relativity. I do not believe there ever was such a time. There might have been a time when only one man did, because he was the only guy who caught on, before he wrote his paper. But after people read the paper a lot of people understood the theory of relativity in some way or other, certainly more than twelve. On the other hand, I think I can safely say that nobody understands quantum mechanics.

What makes relatively a theory which was immediately understood and quantum mechanics a theory which "nobody understands"? Rational reconstructions of scientific revolutions typically attribute theoretical changes to a failure of a specific rule or law. According to this picture, a revolutionary change in theoretical physics occurs when an old and cherished physical principle is abandoned in the face of some new experience, and is replaced by a new and fresh physical principle. Indeed special relativity falls nicely under this description. In this case we can depict a well defined physical principle which "carries the burden of the revolution," namely that of the constancy of the velocity of light in every inertial reference frame. Once we accept this principle -- which is a big step -- then everything follows. The principle of relativity, when added to the classical assumptions regarding isotropy and homogeniety of spacetime, suffices to derive special relativity in all its details. This is the approach which is adopted by many textbooks.

The quantum revolution seems to resist a description in similar terms. What *physical* law, or set of laws can we depict, which will explain the nature of the theoretical change? The first candidate that comes to mind is the uncertainty principle. In its simplest (and, as we shall see later, rather simplistic) formulation it reads as follows:

The operational uncertainty principle: The measurement of the position of a particle disturbs its momentum value and a measurement of its momentum disturbs and introduces uncertainty with respect to its position. The magnitude of the disturbances cannot be reduced below a given fixed value. If Δx, Δp are the standard deviations in the position and momentum respectively, then $\Delta p \Delta x \geq \hbar/2$.

In his famous "γ-ray microscope" analysis, Heisenberg seemed to have had someting like the above "operational" formulation in mind. As stated, this principle is entirely compatible with classical ideas. The "disturbances" mentioned in the formulation do not explain why we need a new type of mechanics. To understand this point, consider an analogous classical case: The measurement of the electrostatic field of a charged sphere at a point p outside it.

The standard "operational definition" of the field is as follows: Bring a positively charged test particle "from infinity" to the point p, measure the electrostatic force, then divide by the charge of the test particle.

But surely the introduction of the test particle causes a disturbance in the value of the original naked field, firstly by causing a redistribution of charges on the surface of the sphere, and secondly by causing a dipol moment effect. Yet no one speaks about "uncertainty in the field value." Even if we add to the above classical description, the non classical assumption that electric charge is quantized, so that the disturbence cannot be made arbitrarily small, no real uncertainty is caused.

There are various reasons for that. Not least among them is the fact that we can calculate, in principle, the precise magnitude of the disturbance and subtract it from the observed value, to obtain the value of the "naked field."

Nothing in the operational formulation of the uncertainty principle prevents us from doing the same thing in the position–momentum case. There may exist a theory which will enable us to make similar calculations in this case. Perhaps we would be able to calculate from the initial conditions, the precise nature and magnitude of a momentum distubance, which occurs during a position measurement? Any such theory would have predictions which transcend those of quantum mechanics.

So the operational formulation of the uncertainty principle falls short of explaining the revolutionary character of quantum theory. Consider therefore an alternative formultion:

The epistemic uncertainty principle: The measurement of a position of a particle disturbes its momentum value and vice versa. The magnitudes of the disturbances cannot be arbitrarily reduced: $\Delta p \Delta x \geq \hbar/2$, and moreover they cannot be *predicted in principle from the values of the initial conditions.*

This formulation perhaps better captures what Heisenberg had in mind (at least in the early days of quantum theory). The trouble is that *this is not a physical law* in any traditional sense of the word. It is a meta-physical law that is, a law concerning all possible physical theories present and future. It maintains that no theory of any sort can predict the values of the "measurement disturbances." There is no way to "test" such a law, for how can we put to test the ability of future theories to make certain predictions?

Another meta-physical way to go about and modify the uncertainty principle, is to attack the problem on the ontological level, rather than the epistemic level:

The ontic uncertainty principle: Physical magnitudes such as "position," "momentum," "energy" and the like, typically associated with physical systems, exist and are well defined only in the context of particular experiments. When a "position" experiment is performed with great accuracy the momentum value is simply undefined and vice versa.

This is the view which is most often cited in textbooks. (This is strange enough, given that one of the main efforts of the analytic school in philosphy was to shift traditional philosophical problems to the epistemic level -- the mind-body problem is a typical case.) Anyway, according to the "ontic uncertainty principle" there really are no "measurement disturbances" at all. The position and momentum of a system are "complementary" magnitudes. We cannot describe the system in terms of both without a contradiction. All we can do is use either one (but then, not the other) when the situation seems fit.

Bohr's own opinion on these matters is hard to discern. It seems to me that a reasonable reconstruction puts Bohr somewhere in between the epistemic and ontic views. The epistemic component in Bohr's position is Kantian: Human beings are bound to think and theorize about physical reality in terms of spatio-temperal relations. Thus, the basic relevant kinematical magnetudes are forced upon physics as a result of the constitution of the human mind. Yet, and here comes the ontological element, Nature would not let itself to be depicted in those terms. Spatio-temporal kinematical magnitudes such as position and velocity and the related dynamical magnetudes such as momentum and energy do not fit the microphysical world. All they yield is a partial and fragmentary description. The reason, presumably, lies with the dual wave-particle character of the elementary constituents of matter. One can't conceive of a "wavicle" in spatio-temoral terms, simply because to us humans the property of being localized (particle), and the property of being extended all over space (wave), seem as an utter contradiction. Since we cannot step outside our minds (or outside language, to use a similar metaphore) we are bound to live with a fragmentary description. Bohr, according to that reconstruction, was a pessimist Kantian.

Be it as it may, one thing is clear, the uncertainty principle cannot be taken as the physical rule which "carries the quantum revolution." In its operational formulation it is simply too weak, and in its metaphyiscal formulations it cannot be taken as a law *of* phyics, but only as a law *about* theoretical physics.

We can look for other candidates, the principle of superposition for example. Again the comparison with the relativity principle is striking. The principle of relativity talks about physical entities and magnitudes: light and velocity. The principle of superposition talks about "state vectors." The question of whether the solutions of the Shrödinger equation represent something real (like wave functions in classical physics), or are mere mathematical codes, is a matter of controversy, analogous to the one which exists with respect to the uncertainty principle. The common view takes $|\psi|^2$, and not ψ, to have a physical meaning. Thus, at least according to the common view, the superposition principle is not *about* anything physical, though it has physical consequences. In any case, the principle of superposition can be taken more as the source of mystery than its solution.

Similar analysis can be made with respect to other candidates. The quantum revolution seems to resist a description, or even reconstruction of the type: "Physical law A has been abandoned and replaced by an alternative physical law B." I take this to be one of the reasons, perhaps the main reason, why "nobody understands quantum mechanics," while everyone seems to be happy with relativity.

Indeed, the gap between the operational formulation, and the other formulations of the uncertainty principle, has been a source of the classical Einstein–Podolsky–Rosen (1935) argument. Consider a particle A at rest, which disintegrates into two particles B and C. We can let the particles B and C move away from one another, and when they are far apart, measure the position of B and the momentum of C. Since by conservation of momentum $p_B + p_C = p_A = 0$ we have $p_B = -p_C$, and we have therefore devised a thought experiment by which we can measure the simultaneous values of the position and momentum of B. As Einstein–Podolsky– Rosen point out, the argument rests on the assumption that the two measurements cannot disturb each other since they are performed when the particles are far apart. This seems to be a reasonable assumption since any known disturbance takes time to propagate. Thus, Einstein–Podolsky–Rosen concluded that the scope of the operational uncertainty principle is limited and the metaphysical formulations are clearly false.

Yet, as is well known, the E.P.R. argument is false. Indeed, we have $\Delta p_C \Delta x_B \geq \hbar/2$, which is very strange indeed. This has been established experimentally with respect to spin or polarization values, though not with respect to position and momentum, and we shall discuss this subject in great detail subsequently. These facts, which were discovered in late the Sixties and in the Seventies, add immense difficulties to any realistic conception of quantum mechanics. If we want to attribute the results of quantum experiments to real measurement disturbances, we have to assume that these disturbances propagate faster than light. All this indicates that we should look for a better understanding of quantum theory elsewhere.

In my view the problem of interpretation of quantum mechanics has two levels: the observational–formal level, and the level of "meaning." I believe that we should therefore take the following steps:

1.   Provide an account for the difference in predicted and observed phenomena between classical and quantum physics on a formal mathematical level.

2.   Attempt to explain the source of the difference. More precisely, to provide a list of possible alternative explanations compatible with the formal analysis mentioned in 1.

The first step is a logical one, since one has to identify the principle or set of principles which explain, in the logical deductive sense, the difference between classical and quantum phenomena. My main effort lies in this direction and the concept which I take to "carry the burden of the revolution" is that of probability, or more precisely that of "correlation".

We shall see that the set of axioms for classical probability entail that frequencies should obey an a–priori set of constraints that are often violated by quantum frequencies. The violation itself has a–priori nothing to do with the principle of locality for it often occurs in cases where spatio–temporal aspects play no role whatever. This I take to be the central experimental fact which distinguishes between classical and quantum physics.

In Chapter 2, which is purely mathematical, I analyze the constriants imposed on correlations by the classical axioms of probability. As we shall see, these constraints are also tightly connected to propositional logic. Chapter 3 is a similar analysis with respect to quantum frequencies, the derivation is based upon the Hilbert space formalism, and E.P.R. type of experiments are discussed in detail.

I do not claim that mine is the only possible approach to the problem. Its advantage lies in its generality, for it provides a good ground to compare a variety of possible interpretations of the meaning of the quantum revolution. Given that we have accomplished step 1 in the process of interpretation, step 2 consists in an attempt to answer the question: What reason is there, if any, for the violation of the classical constraints? There are various possible answers to this question. None of them is implied or even suggested by the formalism of quantum theory itself, and all are compatible with it. I shall discuss a few:

(i)   Nothing causes the violation of the classical constraints. This is the antirealist view, in the spirit of Hume, which claims that events need not have causes, they simply occur. In a world devoid of causes, frequencies of events need not obey the classical constraints.

(ii)   A similar answer, in the tradition of Bohr, does not claim that quantum events do not have causes. It rather claims that the results of experiments need not be located in the physical properties of the system under observation alone, but rather in the properties of the measurement process as a whole.

(iii)   That the violation of classical constraints on correlations is caused by measurement disturbances. This is the classical hidden variable approach. We shall see in Chapters 3 and 5 that this view is committed to the violation of locality.

(iv)   That the violation of classical constraints on correlations is a result of a non classical logical structure which underlies quantum properties. Quantum logic is discussed in Chapter 4. We shall see that when interpreted realistically, quantum logic turns into a classical hidden variable theory in disguise, and is also committed to a violation of locality.

(v)   That the violation of classical constraints occurs because of "wild" nonmeasurable distributions of physical properties. This approach is developed in Chapter 5. It is shown that under such assumptions perfectly local hidden variable theories become possible. In this framework the violation of the classical constraints by frequencies is the rule rather than the exception.

I do not claim that this is an exhaustive list of alternatives. These however do seem to me to be the only *reasonable* approaches which are presently available. I make no effort to decide among them, simply because I feel that no rational guideline for such a decision currently exists. Any choice, at this stage, depends on metaphysical inclinations, which lie beyond the scope of either physics or logic. I frankly doubt whether a more educated decision will ever be possible. These, and other related philosophical issues, are discussed in the last chapter.

2. Classical Correlation Polytopes

and Propositional Logic

2.1 Introduction

Consider an urn containing N balls of different colors sizes and compositions. Let p_1 denote the proportion of red balls in the urn,

p_1 = (number of red balls) $\times$ N^{-1}, p_2 denotes the proportion of wooden balls in the urn, and p_{12} denotes the proportion of balls which are both red and wooden. Clearly

$$0 \leq p_{12} \leq p_2 \leq 1 \qquad\qquad 0 \leq p_{12} \leq p_1 \leq 1 \qquad\qquad (2\text{-}1)$$

p_1, p_2, p_{12}, are in fact probabilities. If the balls in the urn are well mixed p_1 is the probability of drawing a red ball at random, and so forth. The numbers p_1, p_2, p_{12}, represent probabilities of two events and their joint respectively only if inequalities (2-1) are satisfied. These inequalities however are insufficient. For example, we cannot have $p_1 = 0.73$, $p_2 = 0.62$ and $p_{12} = 0.31$. The reason is that the probability of drawing a ball which is either red or wooden is $p_1 + p_2 - p_{12}$ and thus we must have

$$p_1 + p_2 - p_{12} \leq 1 \qquad\qquad (2\text{-}2)$$

An inequality which is violated by the above numbers.

Inequalities (2-1) and (2-2) together are necessary and sufficient for the numbers p_1, p_2, p_{12} to represent probabilities of two events and their joint respectively. (The proofs of this and some less trival claims made in this introduction are postponed to the next section.)

These simple observations have a direct geometric interpretation. Consider the three dimensional real space, and in it the set of all vectors of the form (p_1, p_2, p_{12}), where p_1, p_2, p_{12} satisfy inequalities (2-1) and (2-2). This set is a closed convex polytope whose vertices are the points (0,0,0), (1,0,0), (0,1,0), (1,1,1), (figure (2-1)).

The vertices represent extreme cases: (0,0,0) is a case where no ball in the urn is red and none is wooden, hence certainly no ball has both properties. (0,1,0) is the case where no ball is red, all balls are wooden and of course none has both properties, and so forth. The above polytope and other more general ones which will follow are called "correlation polytopes."

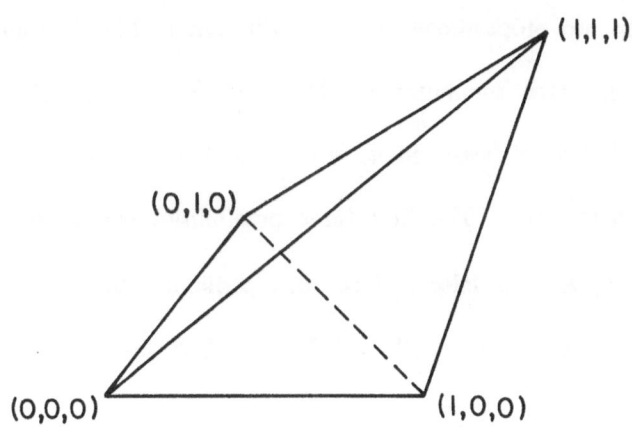

Figure (2–1)

Every convex polytope in a Euclidean space has a dual description: Either in terms of its vertices or in terms of its facets. Under the first description, a given vector is an element of the polytope if and only if it can be represented as a convex combination of the vertices. Under the second description, a vector is an element of the polytope if and only if its coordinates satisfy a set of linear inequalities which represent the supporting hyperplanes of the polytope. The existence of such a dual descripton for every polytope is known as the Weyl–Minkowski theorem. In the specific case of the correlation polytope in Fig. (2–1) our starting point has been the second type of description. We have derived inequalities (2–1) and (2–2) by considering a–priori constraints on proportions of properties. Subsequently we have found the vertices. If we adopt the "subjectivist" approach to probability we can obtain the same result in a reversed order. Consider two propositions a_1 – "It will rain in Madrid tommorrow," and a_2 – "It will rain in Paris tommorow." There are four possibilities: That a_1 and a_2 are both false, that a_1 is false and a_2 is true, that a_1 is true and a_2 false, and that a_1 and a_2 are both true. The first three possibilities correspond to a case where the conjunction $a_1 \wedge a_2$ is false and the last possibility to a case where $a_1 \wedge a_2$ is true. If we let 1 stand for "true" and 0 for "false" these four possibilities can be summed up in the following truth table:

a_1	a_2	$a_1 \wedge a_2$
0	0	0
0	1	0
1	0	0
1	1	1

The rows in this table, if looked at as vectors in a three dimensional Euclidean space, are just the vertices of our polytope.

Suppose that we were to bet on each one of the four possibilities, then we would chose non negative numbers λ_1, λ_2, λ_3, λ_4 where λ_1 is the probability we assign to the event "a_1 is false and a_2 is false," that is, "it will not rain in Madrid nor in Paris tommorrow." λ_3 is the probability we assign to the event "it will rain in Madrid but not in Paris tommorrow" and so forth. Since there are only four possibilities and since they are mutually incompatible we must have $\lambda_1 + \lambda_2 + \lambda_3 + \lambda_4 = 1$. Consider the vector:

$$(p_1, p_2, p_{12}) = \lambda_1(0,0,0) + \lambda_2(0,1,0) + \lambda_3(1,0,0) + \lambda_4(1,1,1) =$$
$$= (\lambda_2 + \lambda_4, \lambda_3 + \lambda_4, \lambda_4)$$

Since it is a convex combination of the vertices it lies in the polytope. $p_1 = \lambda_3 + \lambda_4$ is the probability we have assigned to the proposition "it will rain in Madrid tommorrow," $p_2 = \lambda_2 + \lambda_4$ the probability of the proposition "it wil rain in Paris tommorrow" and $p_{12} = \lambda_4$ is the probability of the conjunction.

In summary we can interpret the vectors in the correlaion polytope in two ways. Under the "objective" view the coordinates p_1, p_2, p_{12} represent proportions or frequencies of two properties and their joint in a given sample. This view is naturally connected to the description of the correlation polytope in terms of linear inequalities. Under the "subjective" view p_1, p_2, p_{12} represent the values of bets on two propositions and their conjunction that is, the weighted average (convex combination) of all truth assignments. This view leads naturally to a description of the correlation polytope in terms of its vertices.

These considerations can be easily generalized. We shall begin with the "subjective" view this time. Consider three propositions a_1, a_2, a_3 and three conjunctions $a_1 \wedge a_2$, $a_1 \wedge a_3$ and $a_2 \wedge a_3$. There are eight possible truth value assignments to three propositions (and thus also to the conjunctions), and these are summed up in the following table.

a_1	a_2	a_3	$a_1 \wedge a_2$	$a_1 \wedge a_3$	$a_2 \wedge a_3$
0	0	0	0	0	0
1	0	0	0	0	0
0	1	0	0	0	0
0	0	1	0	0	0
1	1	0	1	0	0
1	0	1	0	1	0
0	1	1	0	0	1
1	1	1	1	1	1

Now consider each row of the table as a vector in a six dimensional space. There are eight such vectors and their closed convex hull in the six dimensional real space is a correlation polytope which I shall call "the Bell–Wigner polytope" for reasons which will be specified later. Denote vectors in the six dimensional space by $(p_1, p_2, p_3, p_{12}, p_{13}, p_{23})$. Such a vector is an element of the Bell–Wigner polytope if and only if it can be represented as a convex combination of the eight vertices.

What about the description in terms of linear inequalities? In order to establish it consider three events with frequencies p_1, p_2, p_3 and correlations p_{12}, p_{13}, p_{23} (p_{12} is the frequency of the first and second event occuring simultaneously and so forth). Obviously each pair out of the three events must satisfy the inequalities for pair of events and their joint so we have for $1 \leq i < j \leq 3$:

$$0 \leq p_{ij} \leq p_i \leq 1, \qquad 0 \leq p_{ij} \leq p_j \leq 1, \qquad p_i + p_j - p_{ij} \leq 1 \qquad (2\text{-}3)$$

But inequalities (2-3) are not sufficient, we should add to them constraints on all three events (and not just the pairs). The probability that the first event, or the second event, or the third event will occur is greater or equal to

$p_1 + p_2 + p_3 - p_{12} - p_{13} - p_{23}$. Hence we must have

$$p_1 + p_2 + p_3 - p_{12} - p_{13} - p_{23} \leqq 1 \qquad (2\text{--}4)$$

Now instead of the first event substitute its complement (so that if the event is "drawing red ball" the complement is "drawing a ball which is not red"). We have to replace p_1 by $1 - p_1$, p_{12} by $p_2 - p_{12}$, and p_{13} by $p_3 - p_{13}$ while p_2, p_3, p_{23} remain intact. Substituting these values into (2--4) we obtain

$$p_1 - p_{12} - p_{13} + p_{23} \geqq 0 \qquad (2\text{--}5)$$

And by similar arguments

$$\begin{aligned} p_2 - p_{12} - p_{23} + p_{13} &\geqq 0 \qquad (2\text{--}6) \\ p_3 - p_{13} - p_{23} + p_{12} &\geqq 0 \end{aligned}$$

Satisfying inequalities (2--3) through (2--6) is a necessary and sufficinet condition for the vector $p = (p_1,\ p_2,\ p_3,\ p_{12},\ p_{13},\ p_{23})$ to be an element of the Bell–Wigner polytope (as we shall prove later).

The rest of this chapter is concerned with a generalization of these ideas to an arbitrary number of events and some (not necessarily all) of their joints. Since the center of my concern is the relations between logic and probability I shall begin with a brief account of some standard results from the formal propositonal calculus. Readers who are familiar with this subject can skip to Section 2.3.

2.2   Survey of the Propostional Calculus

The propostional calculus is the formal study of propositions, formed from a set of atomic propositions, by means of the logical connectives "not", "and", "or" and "entails". Let $A = \{a_1, a_2, ..., a_n, ...\}$ be a countable set whose elements are called *atomic propositions*. The set of all (well formed) propositions W is defined from A by induction as follows:

(i)   Every atomic proposition $a_i \in A$ is a proposition.

(ii)   If b is a proposition then ($\sim$b) (read: "not b"), is a proposition.

(iii)   If b, c are propositions so are (b $\wedge$ c) (read: "b and c") (b $\vee$ c) (read: "b or c") and (b $\to$ c) (read: b entails c)

Thus every well formed proposition $b \in W$ is a finite string of symbols, each symbol is either one of the atomic propositions a_i, one of the connective symbols $\sim$, $\wedge$, $\vee$, $\to$ or a parantheses symbol (,). Moreover, if b is a non atomic proposition it can be *uniqely* presented as b = ($\sim$b'), or b = (b' $\wedge$ b"), or b = (b' $\vee$ b"), or b = (b' $\to$ b") where b', b" are well formed propositions.

Given a proposition $b = b(a_1, ..., a_n)$ whose atomic constituents are $a_1, a_2, ..., a_n$ a *truth function* is a function t: $\{a_1, a_2, ..., a_n\} \to \{0,1\}$. The *truth value* of the proposition b is defined by induction on the length of formula as follows:

$$t(\sim b) = 1 - t(b)$$
$$t(b \wedge c) = t(b)t(c) \qquad\qquad\qquad (2\text{-}7)$$
$$t(b \vee c) = t(b) + t(c) - t(b)t(c)$$
$$t(b \rightarrow c) = 1 - t(b) + t(b)t(c)$$

For example if $b = ((a_1 \wedge a_2) \rightarrow (\sim a_3))$ and $t: \{a_1, a_2, a_3\} \rightarrow \{0,1\}$ then $t(b) = 1 - t(a_1)t(a_2)t(a_3)$. From this definition it follws that the truth value of every proposition b is fixed once a truth function on its atomic constitutents is chosen. If $b = b(a_1, ..., a_n)$ there are 2^n truth functions $t: \{a_1, ..., a_n\} \rightarrow \{0,1\}$ and each such function determines a value $t(b) \in \{0,1\}$.

If $t(b) = 1$ for every truth function defined on the constitutents of b, then b is called *a logical tautology*. Examples of tautologies are: $b = (c \vee (\sim c))$ for every proposition c, $b = (((\sim c_1)) \rightarrow (\sim c_2)) \rightarrow (((\sim c_1) \rightarrow c_2) \rightarrow c_1))$ for all propositions c_1, c_2 and $b = (((c_1 \wedge c_2) \vee c_3) \rightarrow ((c_1 \vee c_3) \wedge (c_2 \vee c_3)))$ for all propositions c_1, c_2, c_3.

If $t(b) = 0$ for every truth function t, then b is called *a logical falsity*.

Two propositions b_1, b_2 are *equivalent* if $t(b_1) = t(b_2)$ for every truth function defined on their atomic constitutents. In this case we shall write $b_1 \equiv b_2$. For example we have $((a_1 \wedge a_2) \wedge a_3) \equiv (a_1 \wedge (a_2 \wedge a_3))$, so the conjunction $\wedge$ is associative and we can simply write $(a_1 \wedge a_2 \wedge a_3)$, without ambiguity. Similar observation holds for the dijunction symbol $\vee$. For an atomic proposition $a_i \in A$ denote $a_i^1 = a_i$ and $a_i^0 = (\sim a_i)$ then we have

Lemma (2-1): Every proposition $\underline{b} = \underline{b(a_1, ..., a_n)}$ is equivalent to a proposition with the normal disjunctive form:

$$b \equiv \bigvee_{i \in N} \bigwedge_{j=1}^{n} a_j^{\varepsilon_{ij}} \tag{2-8}$$

where $N \subseteq \{1, 2, ..., 2^n\}$ is a set of indices and $\varepsilon_{ij} \in \{0,1\}$.

Proof: Let $b = b(a_1, ..., a_n)$ and let t_i $i = 1, 2, ..., 2^n$ be all the truth functions on $\{a_1, ..., a_n\}$ parameterized in some arbitrary order. Then $t_i(a_j^{t_i(a_j)}) = 1$, to see that consider the case $t_i(a_j) = 1$, then $a_j^{t_i(a_j)} = a_j$ and thus $t_i(a_j^{t_i(a_j)}) = 1$. Similarily if $t_i(a_j) = 0$ then $a_j^{t_i(a_j)} = (\sim a_j)$, hence $t_i(a_j^{t_i(a_j)}) = t_i(\sim a_j) = 1 - t_i(a_j) = 1$. Now put $c_i = a_1^{t_i(a_1)} \wedge a_2^{t_i(a_2)} \wedge ... \wedge a_n^{t_i(a_n)}$ then $t_i(c_i) = 1$. Let

$N = \{i \mid 1 \leq i \leq 2^n, t_i(b) = 1\}$. If N is not empty we have for

$d = \vee_{i \in N} c_i$, $t_i(d) = 1$ if and only if $t_i(b) = 1$ hence

$$d = \bigvee_{i \in N} \bigwedge_{j=1}^{n} a_j^{t_i(a_j)}$$

is equivalent to b. The claim is thus proved for $\varepsilon_{ij} = t_i(a_j)$. If $N = \phi$ we can always write

$$b \equiv \bigvee_{i=1}^{2^n} \bigwedge_{j=1}^{n} a_j^{\delta_{ij}}$$

where $\delta_{ij} = 1 - t_i(a_j)$.

Corrolary (2-2): Every proposition $b \in W$ with atomic constituents $a_1, ..., a_n$ is equivalent to a proposition of the conjuctive normal form

$$b \equiv \bigwedge_{i \in M} \bigvee_{j=1}^{n} a_j^{\delta_{ij}} \tag{2-9}$$

where $M \subseteq \{1, 2, ..., 2^n\}$ is an index set and $\delta_{ij} \in \{0,1\}$.

The proof is left as an excercise for the reader.

2.3 Correlation Polytopes

Let S be a set of pairs of numbers from $\{1,2, ..., n\}$ that is,

$S \subseteq \{\{i,j\} \mid 1 \leq i < j \leq n\}$. We shall denote the cardinality of S by $|S|$. Clearly

$|S| \leq \frac{1}{2}n(n-1)$. Let R(n,S) denote the real space of all functions

f: $\{1,2, ..., n\} \cup S \rightarrow R$. We shall denote vectors in R(n,S) by

$f = (f_1, f_2, ..., f_n, ..., f_{ij}, ...)$ where the numbers f_{ij} appear in a lexicographic order

on the i,j's. Clearly dim R(n,S) = n + $|S|$. Let $\{0,1\}^n$ be the set of all n-tupples of zeroes and one's. We shall denote elements of $\{0,1\}^n$ by $\varepsilon = (\varepsilon_1, ..., \varepsilon_n)$ where $\varepsilon_i \in \{0,1\}$. For each $\varepsilon \in \{0,1\}^n$ let u^ε be the following vector in $R(n,S)$:

$$u_i^\varepsilon = \varepsilon_i \quad 1 \leq i \leq n ; \qquad u_{ij}^\varepsilon = \varepsilon_i\varepsilon_j \quad \{i,j\} \in S \qquad (2\text{-}10)$$

Definition (2-1): The correlation polytope $c(n,S)$ is the closed convex hull in $R(n,S)$ of the 2^n vectors u^ε, $\varepsilon \in \{0,1\}^n$

The vectors u^ε have a straightforward interpretation. Let $a_1, ..., a_n$ be atomic propositions. Each element $\varepsilon = (\varepsilon_1, ..., \varepsilon_n) \in \{0,1\}^n$ is nothing but a truth function on $a_1, ... a_n$ given by $t(a_i) = \varepsilon_i$. Hence $t(a_i) = u_i^\varepsilon$ for $1 \leq i \leq n$ and

$t(a_i \wedge a_j) = u_{ij}^\varepsilon = \varepsilon_i\varepsilon_j$ for $\{i,j\} \in S$. It follows that each u^ε is a row in the truth table for $a_1, ..., a_n$ and the conjunctions $a_i \wedge a_j$, $\{i,j\} \in S$ and every vector in the polytope $c(n,S)$ is a weighted average (convex combination) of these truth values. The following theorem justifies the name "correlation polytope" for $c(n,S)$:

Theorem (2-3): Let $p = (p_1, ..., p_n, ..., p_{ij}, ...)$ be a vector in $R(n,S)$. Then $p \in c(n,S)$ if and only if there is a probability space (X,Σ,μ) and (not necessarily distinct) events $A_1, A_2, ..., A_n \in \Sigma$ such that:

$$p_i = \mu(A_i) \quad 1 \leq i \leq n \qquad p_{ij} = \mu(A_i \cap A_j) \quad \{i,j\} \in S \qquad (2\text{-}11)$$

Proof: Suppose there is a probability space (X, Σ, μ) and events $A_1, \ldots, A_n \in \Sigma$ such that equation (2–11) is satisfied. We shall show

$p = (p_1, \ldots, p_n, \ldots, p_{ij}, \ldots) \in c(n, S)$. For an arbitrary event $B \in \Sigma$ let $B^1 = B$ and

$B^0 = \bar{B} = X \backslash B$. For $\varepsilon = (\varepsilon_1, \ldots, \varepsilon_n) \in \{0,1\}^n$ let $A(\varepsilon) = A_1^{\varepsilon_1} \cap A_2^{\varepsilon_2} \cap \ldots \cap A_n^{\varepsilon_n}$.

Then $A(\varepsilon) \cap A(\varepsilon') = \phi$ for $\varepsilon \neq \varepsilon'$, $\underset{\varepsilon \in \{0,1\}^n}{\cup} A(\varepsilon) = X$ and $A_i = \underset{\{\varepsilon | \varepsilon_i = 1\}}{\cup} A(\varepsilon)$. Put

$\lambda(\varepsilon) = \mu(A(\varepsilon))$ then $\lambda(\varepsilon) \geq 0$, $\underset{\varepsilon \in \{0,1\}^n}{\Sigma} \lambda(\varepsilon) = 1$ and

$p_i = \mu(A_i) = \underset{\{\varepsilon | \varepsilon_i = 1\}}{\Sigma} \lambda(\varepsilon) = \underset{\varepsilon \in \{0,1\}^n}{\Sigma} \lambda(\varepsilon) \varepsilon_i$ also

$p_{ij} = \mu(A_i \cap A_j) = \underset{\{\varepsilon | \varepsilon_i = \varepsilon_j = 1\}}{\Sigma} \lambda(\varepsilon) = \underset{\varepsilon \in \{0,1\}^n}{\Sigma} \lambda(\varepsilon) \varepsilon_i \varepsilon_j$. Hence $p = \underset{\varepsilon \in \{0,1\}^n}{\Sigma} \lambda(\varepsilon) u^\varepsilon$ and thus

$p \in c(n, S)$ is a convex combination of the vertices u^ε, $\varepsilon \in \{0,1\}^n$.

Conversely suppose $p \in c(n, S)$ then, by definition there exist numbers $\lambda(\varepsilon) \geq 0$

such that $\underset{\varepsilon \in \{0,1\}^n}{\Sigma} \lambda(\varepsilon) = 1$ and $p = \underset{\varepsilon \in \{0,1\}^n}{\Sigma} \lambda(\varepsilon) u^\varepsilon$. Let $X = \{0,1\}^n$, let Σ be the

power set of X and let μ be defined by $\mu(B) = \underset{\varepsilon \in B}{\Sigma} \lambda(\varepsilon)$ for $B \subseteq X$. Let

$A_i = \{\varepsilon | \varepsilon_i = 1\}$ then $\mu(A_i) = \underset{\varepsilon \in \{0,1\}^n}{\Sigma} \lambda(\varepsilon) \varepsilon_i = \underset{\varepsilon \in \{0,1\}^n}{\Sigma} \lambda(\varepsilon) u_i^\varepsilon = p_i$ and

$\mu(A_i \cap A_j) = \underset{\varepsilon \in \{0,1\}^n}{\Sigma} \lambda(\varepsilon) \varepsilon_i \varepsilon_j = \underset{\varepsilon \in \{0,1\}^n}{\Sigma} \lambda(\varepsilon) u_{ij}^\varepsilon = p_{ij}$

and hence the claim follows.

The polytope $c(n, S)$ has non empty interior in $R(n, S)$. To see that consider

the vectors u^ε for $\varepsilon = (0, \ldots, 0, \overset{i}{1}, 0, \ldots, 0)$ $1 \leq i \leq n$ and

$\varepsilon = (0, \ldots, 0, \overset{i}{1}, 0, \ldots, 0, \overset{j}{1}, 0, \ldots, 0)$ for $\{i,j\} \in S$. There are $n + |S|$ such vectors

and it is clear that they are linearily independent in $R(n, S)$. Since dim $R(n,S) = n + |S|$ we have proved that among the vertices of $c(n,S)$ there is a basis for $R(n,S)$, hence $c(n,S)$ has non empty interior.

In this way we have defined the general correlation polytope by its vertices, that is, we have adopted a "subjectivist" point of view. What about the "objective" characterization in terms of linear inequalities? This task turns out to be extremely complex. We shall take a few specific examples in the following sections, examples which will serve us in the future. In Section 2.6 we shall derive some (not all) inequalities for the general case and in Section 2.7 we shall prove formally that describing all inequalities for $c(n,S)$ in the general case is practically impossible, since it will probably require too much (i.e. exponential) comptutation time.

The simple case is obtained for $n = 2$ and $S = \{\{1,2\}\}$. The inequalities $0 \leq p_{12} \leq p_1 \leq 1$, $0 \leq p_{12} \leq p_2 \leq 1$ and $p_1 + p_2 - p_{12} \leq 1$ are sufficient, since if these inequalities are satisfied we can always write

$$p = (p_1, p_2, p_{12}) = (1 - p_1 - p_2 + p_{12})(0,0,0) + (p_1 - p_{12})(1,0,0) +$$
$$(p_2 - p_{12})(0,1,0) + p_{12}(1,1,1)$$

and $(0,0,0)$, $(1,0,0)$, $(0,1,0)$, $(1,1,1)$ are just the vertices of the correlation polytope $c(n,S)$ in that case.

2.4 The Bell-Wigner Polytope

The Bell Wigner Polytope is the polytope $c(n,S)$ for $n = 3$ and

$S = \{\{1,2\}, \{1,3\}, \{2,3\}\}$. In that case dim $R(n,S) = 6$ and $c(n,S)$ has eight

vertices u^ε, for $\varepsilon = (0,0,0,), (1,0,0), (0,1,0), (0,0,1), (1,1,0), (1,0,1), (0,1,1), (1,1,1)$.

We shall prove that $p = (p_1, p_2, p_3, p_{12}, p_{13}, p_{23}) \in c(n,S)$ if and only if the

following inequalities are satisfied:

$$
\begin{array}{lll}
0 \leq p_{ij} \leq p_j \leq 1 \quad 0 \leq p_{ij} \leq p_i \leq 1 & 1 \leq i < j \leq 3 & \\
p_i + p_j - p_{ij} \leq 1 & 1 \leq i < j \leq 3 & \\
p_1 + p_2 + p_3 - p_{12} - p_{13} - p_{23} \leq 1 & & (2\text{--}12) \\
p_1 - p_{12} - p_{13} + p_{23} \geq 0 & & \\
p_2 - p_{12} - p_{23} + p_{13} \geq 0 & & \\
p_3 - p_{13} - p_{23} + p_{12} \geq 0 & &
\end{array}
$$

The last four inequalities are called "Bell inequalities," and they are violated in

certain quantum experiments. This fact will be analyzed in great detail in Chapter

3.

Theorem (2-4): p is a vector in the Bell Wigner polytope if and only if

inequalities (2-12) are satisfied

Proof: Necessity has already been proved in the introductory section of this

chapter. A more direct proof of necessity, which avoids the probablistic

characterization of the Bell-Wigner polytope consists in showing that every vertex

u^ε of c(3,S) satisfies inequalities (2–12), and therefore every convex combination of these eight vertices satisfies them as well.

As for the "if" part, suppose p satisfies inequalities (2–12). Let η be any number which satisfies:

$$\eta \leq \min\{p_{12}, p_{13}, p_{23}, 1 - (p_1 + p_2 + p_3 - p_{12} - p_{13} - p_{23})\}$$

$$\eta \geq \max\{0, -p_1 + p_{12} + p_{13}, -p_2 + p_{12} + p_{23}, -p_3 + p_{13} + p_{23})\} \tag{2–13}$$

From inequalities (2–12) it is clear that this definition is consistent, that is, every number which appears in the set on top is greater or equal to any number which appears in the set on bottom.

Now define numbers $\lambda(\varepsilon) = \lambda(\varepsilon_1, \varepsilon_2, \varepsilon_3)$ for each $\varepsilon = (\varepsilon_1, \varepsilon_2, \varepsilon_3) \in \{0,1\}^3$, in the following way:

$$
\begin{aligned}
\lambda(0,0,0) &= 1 - (p_1 + p_2 + p_3 - p_{12} - p_{13} - p_{23}) - \eta \\
\lambda(1,0,0) &= \eta + (p_1 - p_{12} - p_{13}) \\
\lambda(0,1,0) &= \eta + (p_2 - p_{12} - p_{23}) \\
\lambda(0,0,1) &= \eta + (p_3 - p_{13} - p_{23}) \\
\lambda(1,1,0) &= p_{12} - \eta \\
\lambda(1,0,1) &= p_{13} - \eta \\
\lambda(0,1,1) &= p_{23} - \eta \\
\lambda(1,1,1) &= \eta
\end{aligned}
\tag{2–14}
$$

From the definition of η in formula (2–13) it is clear that $\lambda(\varepsilon) \geq 0$ also

$\sum_{\varepsilon \in \{0,1\}^n} \lambda(\varepsilon) = 1$ as can easily be verified.

Now $\sum\limits_{\varepsilon \in \{0,1\}^3} \lambda(\varepsilon)u^\varepsilon = p$. For example, we have

$\lambda(0,1,1) + \lambda(1,1,1) = p_{13} - \eta + \eta = p_{13}$ as required. Another example

$\lambda(0,1,0) + \lambda(1,1,0) + \lambda(0,1,1) + \lambda(1,1,1) = \eta + (p_2 - p_{12} - p_{23}) + (p_{12} - \eta) +$

$(p_{23} - \eta) + \eta = p_2$ as required. The other identities actually follow from

symmetry, or can be varified directly.

2.5 The Clauser–Horne Polytope

The Clauser Horne Polytope is the polytope c(n,S) for n=4 and

$S = \{\{1,3\}, \{1,4\}, \{2,3\}, \{2,4\}\}$. Note that in this case we consider only four out

of the six pairs. We have dim $R(4,S) = 8$ and the Clauser–Horne polytope c(4,S)

has 16 vertices u^ε corresponding to the elements $\varepsilon = (\varepsilon_1, \varepsilon_2, \varepsilon_3, \varepsilon_4)$ of $\{0,1\}^4$. The

inequalities for the Clauser–Horne polytope are:

$$
\begin{aligned}
&0 \le p_{ij} \le p_i \le 1, \quad 0 \le p_{ij} \le p_j \le 1 \quad i = 1,2 \quad j = 3,4 \\
&p_i + p_j - p_{ij} \le 1 \qquad\qquad\qquad\qquad\quad i = 1,2 \quad j = 3,4
\end{aligned}
$$

$$
\begin{aligned}
-1 &\le p_{13} + p_{14} + p_{24} - p_{23} - p_1 - p_4 \le 0 \\
-1 &\le p_{23} + p_{24} + p_{14} - p_{13} - p_2 - p_4 \le 0 \\
-1 &\le p_{14} + p_{13} + p_{23} - p_{24} - p_1 - p_3 \le 0 \\
-1 &\le p_{24} + p_{23} + p_{13} - p_{14} - p_2 - p_3 \le 0
\end{aligned}
\qquad (2\text{-}15)
$$

The last four inequalities are called the Clauser–Horne inequalities and, again, they

play a role in quantum statistics which will be explained in the next chapter.

Theorem (2-5): p is an element of the Clauser-Horne polytope if and only if it satisfies inequalites (2-15)

Proof: Necessity is easy to prove directly. Take any vertex

$$u^{\varepsilon} = (\varepsilon_1, \; \varepsilon_2, \; \varepsilon_3, \; \varepsilon_4, \; \varepsilon_1\varepsilon_3, \; \varepsilon_1\varepsilon_4, \; \varepsilon_2\varepsilon_3, \; \varepsilon_2\varepsilon_4), \; \varepsilon \in \{0,1\}^4 \text{ and prove that } u^{\varepsilon} \text{ satisfy}$$

inequalities (2-15). As for sufficiency choose a number γ which satisfies

$$\gamma \leq \min\{p_1, \; p_2, \; p_1 - p_{13} + p_{23}, \; p_2 - p_{23} + p_{13}, \; p_1 - p_{14} + p_{24}, \\ p_2 - p_{24} + p_{14}, \; 1\}$$

$$\gamma \geq \max\{p_1 + p_2 - 1, \; p_{13} + p_{23} - p_3, \; p_{14} + p_{24} - p_4, \\ p_1 + p_2 + p_3 - p_{13} - p_{23} - 1, \; p_1 + p_2 + p_4 - p_{14} - p_{24} - 1, \; 0\}$$

From inequalities (2-15) it follows that the definition of γ is consistent, that is, every number in the top set is greater or equal to any number in the bottom set. Now define a vector $p' = (p_1', \; p_2', \; p_3', \; p_{12}', \; p_{13}', \; p_{23}')$ in the space $R(3,S')$ where $S' = \{\{1,2\}, \; \{1,3\}, \; \{2,3\}\}$ by $p_1' = p_1, \; p_2' = p_2, \; p_3' = p_3, \; p_{12}' = \gamma, \; p_{13}' = p_{13}, \; p_{23}' = p_{23}$, then p' is in the Bell Wigner polytope $c(3,S')$. In order to see that we have to verify that the Bell Wigner inequalities (2-12) are satisfied for p' and indeed they are, for example $p_1' + p_2' - p_{12}' = p_1 + p_2 - \gamma \leq 1$ since $\gamma \geq p_1 - p_2 - 1$ also $p_1' - p_{12}' - p_{13}' + p_{23}' = p_1 - \gamma - p_{13} + p_{23} \geq 0$. Since $\gamma \leq p_1 - p_{13} + p_{23}$ and so forth. Hence by theorem (2-5) there exist $\lambda'(\varepsilon) = \lambda'(\varepsilon_1, \; \varepsilon_2, \; \varepsilon_3) \geq 0$ such that $\sum\limits_{\varepsilon \in \{0,1\}^3} \lambda'(\varepsilon) = 1, \; p' = \sum\limits_{\varepsilon \in \{0,1\}^3} \lambda'(\varepsilon)u^{\varepsilon}$.

Similarly let $p'' = (p_1'', \; p_2'', \; p_3'', \; p_{12}'' \; p_{13}'' \; p_{23}'')$ be defined by $p_1'' = p_1, \; p_2'' = p_2, \; p_3'' = p_4, \; p_{12}'' = \gamma, \; p_{13}'' = p_{14}, \; p_{23}'' = p_{24}$ and again by the very same argument p'' is an element of the Bell–Wigner polytope and there are numbers

$\lambda''(\varepsilon) = \lambda''(\varepsilon_1, \varepsilon_2, \varepsilon_3) \geq 0$ for all $\varepsilon \in \{0,1\}^3$ such that $\sum\limits_{\varepsilon \in \{0,1\}^3} \lambda''(\varepsilon) = 1$ and

$\sum\limits_{\varepsilon \in \{0,1\}^3} \lambda''(\varepsilon) u^\varepsilon = p''.$

Now for every $\varepsilon = (\varepsilon_1, \varepsilon_2, \varepsilon_3, \varepsilon_4) \in \{0,1\}^4$ define

$$\lambda(\varepsilon) = \lambda(\varepsilon_1, \varepsilon_2, \varepsilon_3, \varepsilon_4) = \frac{\lambda'(\varepsilon_1, \varepsilon_2, \varepsilon_3)\, \lambda''(\varepsilon_1, \varepsilon_2, \varepsilon_4)}{\lambda'(\varepsilon_1, \varepsilon_2, 0) + \lambda'(\varepsilon_1, \varepsilon_2, 1)} \qquad (2\text{--}16)$$

When the denominator is non zero.

Note that since $p_1' = p_1'' = p_1$, $p_2' = p_2'' = p_2$, $p_{12}' = p_{12}'' = \gamma$ we have

$\lambda'(\varepsilon_1, \varepsilon_2, 0) + \lambda'(\varepsilon_1, \varepsilon_2, 1) = \lambda''(\varepsilon_1, \varepsilon_2, 0) + \lambda''(\varepsilon_1, \varepsilon_2, 1)$, so the denominator in

(2–16) would have been the same had we chosen λ'' instead of λ'.

In case $\lambda'(\varepsilon_1, \varepsilon_2, 0) + \lambda'(\varepsilon_1, \varepsilon_2, 1) = 0$ let $\lambda(\varepsilon_1, \varepsilon_2, \varepsilon_3, \varepsilon_4) = 0$ for all

$\varepsilon_3, \varepsilon_4 \in \{0,1\}$. It is clear that $\lambda(\varepsilon) \geq 0$ also:

$$\sum\limits_{\varepsilon \in \{0,1\}^4} \lambda(\varepsilon) = \sum\limits_{\varepsilon_1 \varepsilon_2} \sum\limits_{\varepsilon_3 \varepsilon_4} \frac{\lambda'(\varepsilon_1, \varepsilon_2, \varepsilon_3)\, \lambda''(\varepsilon_1, \varepsilon_2, \varepsilon_4)}{\lambda'(\varepsilon_1, \varepsilon_2, 0) + \lambda'(\varepsilon_1, \varepsilon_2, 1)} =$$

$$= \sum\limits_{\varepsilon_1 \varepsilon_2} \frac{[\lambda'(\varepsilon_1, \varepsilon_2, 0) + \lambda'(\varepsilon_1, \varepsilon_2, 1)]\,[\lambda''(\varepsilon_1, \varepsilon_2, 0) + \lambda''(\varepsilon_1, \varepsilon_2, 1)]}{\lambda'(\varepsilon_1, \varepsilon_2, 0) + \lambda'(\varepsilon_1, \varepsilon_2, 1)} =$$

$$= \sum\limits_{\varepsilon_1 \varepsilon_2} [\lambda''(\varepsilon_1, \varepsilon_2, 0) + \lambda''(\varepsilon_1, \varepsilon_2, 1)] = \sum\limits_{\varepsilon \in \{0,1\}^3} \lambda''(\varepsilon) = 1$$

We have $\Sigma\ \lambda(\varepsilon)\varepsilon_i\ =\ p_i$ for $i\ =\ 1,2,3,4$ and $\Sigma\ \lambda(\varepsilon)\varepsilon_i\varepsilon_j\ =\ p_{ij}$ for $i\ =\ 1,2, j\ =\ 3,4$.

As can easily be verified. For example

$$\lambda(1,0,0,1)\ +\ \lambda(1,0,1,1)\ +\ \lambda(1,1,0,1)\ +\ \lambda(1,1,1,1)\ =$$

$$=\ \frac{\lambda'(1,0,0)\lambda''(1,0,1)+\lambda'(1,0,1)\lambda''(1,0,1)}{\lambda'(1,0,0)\ +\ \lambda'(1,0,1)}\ +\ \frac{\lambda'(1,1,0)\lambda''(1,1,1)+\lambda'(1,1,1)\lambda''(1,1,1)}{\lambda'(1,1,0)\ +\ \lambda'(1,1,1)}$$

$$=\ \lambda''(1,0,1)\ +\ \lambda''(1,1,1)\ =\ p''_{13}\ =\ p_{14} \text{ as required.}$$

2.6 Symmetries and Some Inequalities of c(n,S)

We shall see in the next section that deriving all facet inequalities of $c(n,S)$ for an arbitrary n and S is a task which is too complex to be carried out in practice. In this section we shall identify symmetries of the polytope $c(n,S)$ and utilize them to derive a large family of inequalities for the general correlation polytope $c(n,S)$. Let $p \in c(n,S)$, by theorem (2-3) there exist a probability space (X,Σ,μ) and events $A_1, A_2, ..., A_n \in \Sigma$ such that $p_i\ =\ \mu(A_i)$ $i\ =\ 1, 2, ..., n$ and $p_{ij}\ =\ \mu(A_i \cap A_j)$ for $\{i,j\} \in S$. Let i be fixed and consider the event $\bar{A}_i\ =\ X\backslash A_i$, the complement of A_i, we have $\mu(\bar{A}_i)\ =\ 1\ -\ \mu(A_i)\ =\ 1\ -\ p_i$ and

$\mu(\bar{A}_i \cap A_j)\ =\ \mu(A_j)\ -\ \mu(A_i \cap A_j)\ =\ p_j\ -\ p_{ij}$, whenever j is such that $\{i,j\} \in S$. Hence define a transformation σ_i on the space $R(n,S)$ as follows:

if $x\ =\ (x_1\ ...\ x_n,\ ...\ x_{ij}\ ...) \in R(n,S)$ then $(\sigma_i x)$ is defined by

$$(\sigma_i x)_i = 1 - x_i$$

$$(\sigma_i x)_j = x_j \qquad\qquad \text{for all } j \neq i$$

$$(\sigma_i x)_{ij} = x_j - x_{ij} \qquad \text{whenever } \{i,j\} \in S$$

$$(\sigma_i x)_{jk} = x_{jk} \qquad\qquad \text{for } \{j,k\} \in S \ j,k \neq i$$

(2-17)

Note that σ_i is *not* a linear transformation (it is an affine transformation). From the above remark it follows that σ_i leaves the polytope $c(n,S)$ invariant. Moreover σ_i exchange the vertices of $c(n,S)$: $\sigma_i u^\varepsilon = u^\delta$ for

$\delta = (\varepsilon_1 \ \ldots \ \varepsilon_{i-1}, \ 1 - \varepsilon_i, \ \varepsilon_{i+1} \ \ldots \ \varepsilon_n)$. It is also clear that σ_i^2 is the identity transformation. The transformations σ_i $i = 1,2, \ldots, n$ thus generate a commutative group isomorphic to $Z_2^{(n)}$. Every element σ of this group can be represented by

$$\sigma = \sigma(\varepsilon) = \prod_{i=1}^{n} \sigma_i^{\varepsilon_i} \text{ where } \varepsilon = (\varepsilon_1, \ldots, \varepsilon_n) \in \{0,1\}^n \text{ (modulo the convention}$$

$\sigma_i^0 = $ identity). If $0 = (0,0 \ \ldots, 0)$ then u^0 is the origin in $R(n,S)$ and $\sigma(\varepsilon)u^0 = u^\varepsilon$, hence any vertex of $c(n,S)$ can be obtained from any other vertex of $c(n,S)$ by an oerpation of the group $Z_2^{(n)}$ as above.

Next let $\pi\colon \{1, 2, \ldots, n\} \to \{1, 2, \ldots, n\}$ be a permutation such that $\{i,j\} \in S$ if and only if $\{\pi(i), \pi(j)\} \in S$. For every such permutation π we shall define a (linear) transformation on $R(n,S)$, which we shall also denote by π, by

$$(\pi x)_i = x_{\pi(i)} \qquad\qquad i = 1,2, \ldots, n$$

$$(\pi x)_{ij} = x_{\pi(i)\pi(j)} \qquad \{i,j\} \in S$$

(2-18)

Then, π leaves the polytope c(n,S) invariant since its effect is to permute the events A_1, ..., A_n. Let $\Pi(n,S)$ denote the group of all transformations of that kind and let G(n,S) be the group generated by both $Z_2^{(n)}$ and $\Pi(n,S)$. The mixed action of complementations and permutations is easy to identify. We have

$$\pi(\prod_{i=1}^{n} \sigma_i^{\varepsilon_i})\pi^{-1} = \prod_{i=1}^{n} \sigma_{\pi(i)}^{\varepsilon_i} \text{ so that } Z_2^{(n)} \text{ is a normal (invariant) subgroup of G(n,S).}$$

In order to see how to apply these symmetries we take first the case where $S = S_n$ is the set of all pairs that is: $S_n = \{\{i,j\}\} \mid 1 \leq i < j \leq n\}$. In this case $\Pi(n,S_n)$ is the group of all permutations. If A_1, ..., A_k are any events in a probability space (X,Σ,μ) then we have

$$\sum_{i=1}^{k} \mu(A_i) - \sum_{1\leq i<j\leq k} \mu(A_i \cap A_j) \leq \mu(A_1 \cup A_2 \cup ... \cup A_k) \leq 1$$

hence if $p \in c(n,S_n)$ we must have

$$\sum_{i=1}^{k} p_i - \sum_{1\leq i<j\leq k} p_{ij} \leq 1 , \qquad 1 \leq k \leq n \qquad\qquad (2\text{--}19)$$

If this is true for p this must be true for σp and πp whenever $\sigma \in Z_2^{(n)}$ and $\pi \in \Pi(n,S)$. Summing up these facts we conclude:

$$\sum_{i\in\alpha} (\sigma p)_i - \sum_{\substack{i<j \\ ij\in\alpha}} (\sigma p)_{ij} \leq 1 \qquad\qquad (2\text{--}20)$$

For all non empty subsets $\alpha \subseteq \{1,2, ..., n\}$ and all elements $\sigma \in Z_2^{(n)}$. To see how to apply these inequalities take $\alpha = \{i\}$ then $p_i \leq 1$ applying σ_i we get $1 - p_i \leq 1$ or $p_i \geq 0$, take $\alpha = \{i,j\}$ then $p_i + p_j - p_{ij} \leq 1$ apply σ_i we get

$(1-p_i) + p_j - (p_j - p_{ij}) \leq 1$ or $p_{ij} \leq p_i$ apply σ_j to obtain $p_{ij} \leq p_j$ apply $\sigma_i\sigma_j$ and obtain $(1-p_i) + (1-p_j) - (1-p_i-p_j+p_{ij}) \leq 1$ or $p_{ij} \geq 0$. Now take $\alpha = \{i,j,k\}$ then

$p_i + p_j + p_k - p_{ij} - p_{ik} - p_{jk} \leq 1$ apply σ_i to obtain $p_i - p_{ij} - p_{ik} + p_{jk} \geq 0$ and these are just the Bell inequalities (2–12) for $\{i,j,k\} = \{1,2,3\}$.

In the case $S \subset S_n$ we can still apply inequality (2–20) for every subset $\alpha \subseteq \{1,2,...,n\}$ for which $\{\{i,j\} \mid i < j \; i,j \in \alpha\} \subset S$

2.7   The Computational Intractability of the Generalized Bell Inequalities*

The deeper relations between propositional logic and probability theory is manifested when we attempt to derive all the inequalities for $c(n,S)$. We shall show that the number and complexity of the inequalities grow so fast that it will probably require exponentially many computation steps (exponent in a power of n, that is) to derive them all.

---

*The contents of this section is unrelated to the rest of the monograph, so it can be skipped in the first reading.

Consider the following problem from porpositional logic: Given a proposition b = b(a₁, ..., aₙ) determine whether b is a tautology or not. A simple algorithm comes immediately to mind. We produce all the 2^n truth functions $t:\{a_1, ..., a_n\} \to \{0,1\}$. For each such function we calculate t(b). If t(b) = 1 for all truth functions t then b is a tautology otherwise it is not a tautology.

This algorithm is simple to describe, the trouble lies with its extreme inefficiency in terms of calculation time. Suppose that the calculation of t(b) for a given fixed t requires a mere 10^{-6} seconds, then for n = 60 it will require ~360 centuries to complete the computation of all the 2^{60} cases! The natural question to ask therefore is whether a better algorithm for this purpose exists. Let us make the question more precise. From (2-7) it follows that the calculation of t(b) for a given fixed t requires a sequence of arithmetical binary operations (addition and multiplication). Let us call each such operation a step in the calculation. Moreover, given b = b(a₁, ..., aₙ) let |b| denote the total number of symbols which appear in b (that is, the number of atomic propositions plus the number of logical symbols and parantheses). Now we can phrase our question more precisely.

Problem (2-6) Does there exist a fixed polynomial Q(x) and an algorithm, such that, given a proposition b = b(a₁, ..., aₙ) as input it decides whether b is a tautology or not in less then Q(|b|) steps?

I did not specify what an "algorithm" means. The formal definition is in terms of Turing machines, but for our purposes an intuitive notion will suffice. An algorithm which satisfies the restriction above is called a "polynomial time algorithm."

The answer to the above problem is unknown. This is the most important and most difficult open problem in theoretical computer science. The reason for its centrality is that a positive answer to this problem will automatically entail the existence of polynomial time algorithms for a vast, literatly hundreds, of other computation problems, some of major technological and scientific importance (more on that in the "notes and remarks" section 2.9). Most researchers in the field believe that no polynomial time algorithm for the tautology problem exists. Any algorithm which decides whether a given proposition is a tautology or not will not be polynomial time (and therefore, by definition will be exponetial time).

What all this has to do with the original question regarding the facets of c(n,S)? To establish the connection let I(n,S) be the set of all linear inequalities which describe the facets of c(n,S). More precisely I(n,S) is a set of n + |S| + 1 dimensional vectors $f = (f_1 \ldots f_n, \ldots f_{ij} \ldots, g)$ and $p \in c(n,S)$ if and only if

$$\sum_{i=1}^{n} f_i p_i + \sum_{1 \leq i < j \leq n} f_{ij} p_{ij} \leq g \text{ for all vectors of I(n,S). I have already mentioned the}$$

fact that the Weyl–Minkowski theorem assures us that the (finite) set I(n,S) exists for all n and S.

There are three senses in which the set I(n,S) can be considered simple. These are given by the following three conditions:

SIMP$_1$: Each inequality in the set I(n,S) is not too complex, that is there exists a fixed polynomial $Q_1(x)$ such that the complexity of every vector $(f_1 \ldots f_n, \ldots f_{ij} \ldots, g) \in$ I(n,S) is less than $Q_1(n + |S|)$. (The complexity of a vector is just the number of symbols which are required for its storage. Remember that every natural number m can be coded in terms of lgm symbols.)

SIMP$_2$: The enumeration of I(n,S) is fast, that is, there exists a fixed polynomial $Q_2(x)$ and an algorithm such that when provided with the number n and the set S as input produces the first vector in I(n,S) in less then $Q_2(n + |S|)$ steps, produces the second vector in I(n,S) in less then $Q_2(n + |S|)$ *additional* steps and then the third inequality... and so forth until the list of all vectors in I(n,S) is exhausted. Note that condition SIMP$_2$ entails SIMP$_1$. Since otherwise {I(n,S)| n = 2,3, ..., S set of pairs} contains vectors of assymptotic exponential complexity (in n, |S|) and we have no hope of producing them, in their turn, in polynomial time.

SIMP$_3$: There exists a fixed polynomial $Q_3(x)$ and an algorithm such that when given n, S and a (rational) vector p $\in$ R(n,S) as input decides whether p $\in$ c(n,S) or not in a number of steps which does not exceed $Q_3(|p|)$.

Condition SIMP$_3$ entails that there is a polynomial time algorithm which decides whether all the inequalities in I(n,S) are simultaneously satisfied by a given vector p. But note that SIMP$_3$ does not necessarily entail SIMP$_2$. We may have some indirect way to decide wheter p $\in$ c(n,S) or not, without explicity deriving all the vectors in I(n,S).

What I shall demonstrate is that if condition $SIMP_3$ is satisfied then there exists a polynomial time algorithm which decides whether a given proposition is a tautology or not. Our proof will also demonstrate that it is highly unlikely that $SIMP_2$ is true. The condition $SIMP_1$ *is* demonstratively valid.

Before going into the details of the proof, I shall restrict the tautology problem to a subclass of propositions without limiting its scope. Let a_1, a_2, ..., a_n ... be atomic propositons, put

$$e(a_i, a_j, a_k) = (a_i \lor a_j \lor a_k) \land (\sim(a_i \land a_j)) \land (\sim (a_i \land a_k)) \land \qquad (2\text{--}21)$$
$$\land (\sim (a_j \land a_k))$$

It is clear that a truth value $t: \{a_i, a_j, a_k\} \rightarrow \{0,1\}$ satisfies $t(e(a_i, a_j, a_k)) = 1$ if and only if one and only one of the values of $t(a_i)$, $t(a_j)$, $t(a_k)$ is "true" that is, if and only if, $t(a_i) + t(a_j) + t(a_k) = 1$. We shall say that b is a *simple* proposition if b is a conjunction of propositions of type e in (2–21) for example

$e(a_1, a_2, a_3) \land e(a_1, a_5, a_7) \land e(a_2, a_3, a_4) \land e(a_3, a_4, a_6)$ is a simple proposition or

$e(a_1, a_2, a_6) \land e(a_1, a_2, a_4) \land e(a_1, a_3, a_4) \land e(a_2, a_3, a_4)$ is simple (and is a logical falsity). Now consider the following

Problem (2–7): Does there exist a polynomial time algorithm such that when given a simple proposition b as input, decides whether b is a logical falsity or not?

It turns out that problem (2–7) is equivalent to the full blown tautology problem (2–6). The answer to problem (2–7) is "yes" iff and only if the answer to problem (2–6) is affirmative.

Let k and $4 \leq m \leq \binom{k}{3} = \frac{1}{6}k(k-1)(k-2)$ be two natural numbers and consider the simple proposition.

$$b = e(a_{\nu_1}, a_{\sigma_1}, a_{\eta_1}) \wedge e(a_{\nu_2}, a_{\sigma_2}, a_{\eta_2}) \wedge \ldots \wedge e(a_{\nu m}, a_{\sigma m}, a_{\eta m}) \qquad (2-22)$$

Where for i = 1,2, ..., m $1 \leq \nu_i < \sigma_i < \eta_i \leq k$. We shall assume that for every natural number $1 \leq \nu \leq k$ there is an index $1 \leq i \leq m$ such that $\nu \in \{\nu_i, \sigma_i, \eta_i\}$.

If $\nu \in \{\nu_i, \sigma_i, \eta_i\}$ we shall say that ν occured in the ith conjunct. Let $l(\nu)$ be the number of conjuncts in which ν occures. By assumption $l(\nu) \geq 1$.

Given a fixed proposition as above set n = m + k + 1 and let S be the following family of pairs $S = S_0 \cup S_1 \cup S_2$ where

$S_0 = \{\{\nu, \sigma\} \mid 1 \leq \nu < \sigma \leq k$ and there exists i such that $\nu, \sigma \in \{\nu_i, \sigma_i, \eta_i\}\}$

$S_1 = \{\{\nu, i\} \mid 1 \leq \nu \leq k, k < i \leq k + m, \nu \in \{\nu_i, \sigma_i, \eta_i\}\}$

$S_2 = \{\{i, k + m + 1\} \mid k + 1 \leq i \leq m\}$

Let J be a natural number and define a vector $p^J \in R(n,S)$ in the following way

Domain	Value
$1 \leq v \leq k$	$p_v^J = 3^{-l(v)}$
$k + 1 \leq i \leq k + m$	$p_i^J = 3^{-l(v_i)} + 3^{-l(\sigma_i)} + 3^{-l(\eta_i)}$
$k + m + 1$	$p_{k+m+1}^J = J\,3^{-m}$
$\{v, \sigma\} \in S_0$	$p_{v\sigma}^J = 0$
$\{v, i\} \in S_1$	$p_{vi}^J = 3^{-l(v_i)}$
$\{i, k + m + 1\} \in S_2$	$p_{i,k+m+1}^J = J\,3^{-m}$

Note that if $J \neq J'$ then p^J and $p^{J'}$ are identical except for the coordiantes $k+m+1$ and $\{i, k+m+1\} \in S_2$. The relationship between the proposition b in (2–22) and the vectors p^J is given by the following two lemmas:

<u>Lemma (2–8):</u>   If there exists $J \geq 0$ such that $p^J \in c(n,S)$ then there exists a truth function $t:\{a_1, ..., a_n\} \to \{0,1\}$ such that $t(b) = 1$

<u>Proof:</u>   Suppose that $p^J \in c(n,S)$ for some $J > 0$ then by theorem (2–3) there exists a probability space (X, Σ, μ) and events $A_1 \ ... \ A_k, B_1 \ ... \ B_m, C \in \Sigma$ such that

$1 \leq v \leq k$ $\qquad \mu(A_v) = p_v^J = 3^{-l(v)}$

$k < i \leq m$ $\qquad \mu(B_i) = p_{k+i}^J = 3^{-l(v_i)} + 3^{-l(\sigma_i)} + 3^{-l(\eta_i)}$

$\qquad\qquad\qquad \mu(C) = p_{k+m+1}^J = J\, 3^{-m}$

$\{v,\sigma\} \in S_0$ $\qquad \mu(A_v \cap A_\sigma) = p_{v\sigma}^J = 0$

$\{v,k+i\} \in S_1$ $\qquad \mu(A_v \cap B_i) = p_{v,k+i}^J = 3^{-l(v)}$

$\{i,\, k+m+1\} \in S_2 \quad \mu(B_i \cap C) = p_{i,k+m+1}^J = J\, 3^{-m}$

In the following equality (inclusion) symbol between events in Σ will denote equality (inclusion) up to sets of μ–measure zero. This makes no difference since we can eleminate subsets of measure zero anyway.

First note that $B_i = A_{v_i} \cup A_{\sigma_i} \cup A_{\eta_i}$; the reason is that by definition $\{v_i,\, k+i\}, \{\sigma_i,\, k+i\}, \{\eta_i,\, k+i\} \in S_1$ for all $i = 1,2 \dots,\ m$ hence $\mu(A_{v_i} \cap B_i) = 3^{-l(v_i)} = \mu(A_{v_i})$ and

$\mu(A_{\sigma_i} \cap B_i) = 3^{-l(\sigma_i)} = \mu(A_{\sigma_i})$ and $\mu(A_{\eta_i} \cap B_i) = 3^{-l(\eta_i)} = \mu(A_{\eta_i})$ and thus

$A_{v_i} \cup A_{\sigma_i} \cup A_{\eta_i} \subseteq B_i$. Also

$$\mu(B_i) = 3^{-l(v_i)} + 3^{-l(\sigma_i)} + 3^{-l(\eta_i)} = \mu(A_{v_i}) + \mu(A_{\sigma_i}) + \mu(A_{\eta_i}) =$$
$$= \mu(A_{v_i} \cup A_{\sigma_i} \cup A_{\eta_i})$$

because A_{ν_i}, A_{σ_i}, A_{η_i} are pairwise disjoint (again up to a set of μ–measure zero which can be eliminated). Therefore we conclude $B_i = A_{\nu_i} \cup A_{\sigma_i} \cup A_{\eta_i}$. But $\mu(B_i \cap C) = J \, 3^{-m} = \mu(C)$ hence $C \subset B_i$ for all $i = 1,2, \dots m$ and therefore:

$$C \subseteq \bigcap_{i=1}^{m} B_i = \bigcap_{i=1}^{m} (A_{\nu_i} \cup A_{\sigma_i} \cup A_{\eta_i})$$

Since $\mu(C) > 0$ we have $\mu(\bigcap_{i=1}^{m} B_i) > 0$ and thus $\bigcap_{i=1}^{m} B_i$ is not empty. Let x be an element of $\bigcap_{i=1}^{m} B_i = \bigcap_{i=1}^{m} (A_{\nu_i} \cup A_{\sigma_i} \cup A_{\eta_i})$. Define a truth function

t: $\{a_1, \dots, a_k\} \to \{0,1\}$ by $t(a_\nu) = 1$ if and only if $x \in A_\nu$. Since

$x \in A_{\nu_i} \cup A_{\sigma_i} \cup A_{\eta_i}$ for all $i = 1, 2, \dots, m$ we have $t(a_{\nu_i}) + t(a_{\sigma_i}) + t(a_{\eta_i}) \geq 1$,

$1 \leq i \leq m$. Since A_{ν_i}, A_{σ_i}, A_{η_i} are pairwise disjoint we have

$t(a_{\nu_i}) + t(a_{\sigma_i}) + t(a_{\eta_i}) = 1$ for $1 \leq i \leq m$ and thus $t[e(a_{\nu_i}, a_{\sigma_i}, a_{\eta_i})] = 1$ $1 \leq i \leq m$

and therefore $t(b) = 1$.

Lemma (2-9): Let J be the number of distinct truth functions

t: $\{a_1, \dots, a_k\} \to \{0,1\}$ for which $t(b) = 1$ (J = 0 in case no such truth function

exists) then $p^J \in c(n,S)$

Proof: Let J be as above, let $X = \{1,2,3\}^m$ be the space of all m– tupples of

numbers taken from $\{1,2,3\}$. Let Σ be the power set of X and μ the uniform

probability measure on X that is, $\mu(\{x\}) = 3^{-m}$ for all $x \in X$. For $1 \leq \nu \leq k$ let

A_ν be defined by

$$A_v = D(v,1) \times D(v,2) \times \ldots \times D(v,m)$$

Where $D(v,i) \subseteq \{1,2,3\}$ is given by the following

$$D(v,i) = \begin{cases} \{1\} & v = v_i \\ \{2\} & v = \sigma_i \\ \{3\} & v = \eta_i \\ \{1,2,3\} & v \notin \{v_i, \sigma_i, \eta_i\} \end{cases}$$

Then clearly $\mu(A_v) = 3^{-l(v)} = p_v^J$. Also if $\{v,\sigma\} \in S_0$ there is i such that

$v,\sigma \in \{v_i, \sigma_i, \eta_i\}$ so that $\mu(A_v \cap A_\sigma) = 0 = p_{v\sigma}^J$. Put $B_i = A_{v_i} \cup A_{\sigma_i} \cup A_{\eta_i}$ then

$\mu(B_i) = 3^{-l(v_i)} + 3^{-l(\sigma_i)} + 3^{-l(\eta_i)} = p_{k+i}^J$. If $\{v, i+k\} \in S$ then $v \in \{v_i, \sigma_i, \eta_i\}$ so

that $\mu(B_i \cap A_v) = 3^{-l(v)} = p'_{v,i+k}$. Put $C = \bigcap_{i=1}^{n} B_i = \bigcap_{i=1}^{n} (A_{v_i} \cup A_{\sigma_i} \cup A_{\eta_i})$.

We shall conclude the proof when we show that $\mu(C) = J \, 3^{-m} = p_{k+m+1}^J$. Let

t: $\{a_1, \ldots, a_k\} \to \{0,1\}$ be such that $t(b) = 1$. Define an element

$x(t) = (x_1(t), x_2(t), \ldots, x_m(t)) \in X = \{1,2,3\}^m$ as follows: $x_i(t) = 1$ if

$t(a_{v_i}) = 1 \; x_i(t) = 2$ if $t(a_{\sigma_i}) = 1$ and $x_i(t) = 3$ if $t(a_{\eta_i}) = 1$. Since

$t(a_{v_i}) + t(a_{\sigma_i}) + t(a_{\eta_i}) = 1$ for all $1 \le i \le m$, the element x is well defined. Also if

$t \ne t'$ we have $x(t) \ne x(t')$ (remember we assumed $l(v) \ge 1$ for all $1 \le v \le k$). I

shall show that $x(t) \in A_v$ if and only if $t(a_v) = 1$. To see this let i be such that

$v \in \{v_i, \sigma_i, \eta_i\}$ then if $v = v_i$ we have $D(v,i) = \{1\}$ and $t(a_v) = 1$ iff $x_i(t) = 1$;

similarily if $v = \sigma_i$ then $D(v,i) = \{2\}$ and $t(a_v) = 1$ iff $x_i(t) = 2$; finally if $v = \eta_i$

then $D(v,i) = \{3\}$ and $t(a_v) = 1$ iff $x_i(t) = 3$. Therefore if $v \in \{v_i, \sigma_i, \eta_i\}$ then

$t(a_v) = 1$ if and only if $x_i(t) \in D(v,i)$. If $v \notin \{v_i, \sigma_i, \eta_i\}$ then $D(v,i) = \{1,2,3\}$ and $x_i(t) \in D(v,i)$, hence (since $l(v) \geq 1$) we obtain $t(a_v) = 1$ if and only if $x(t) \in D(v,1)x \dots x D(v,m) = A_v$.

It follows that $t(b) = 1$ iff $t(a_{v_i}) + t(a_{\sigma_i}) + t(a_{\eta_i}) = 1$ iff

$x(t) \in \overset{n}{\underset{i=1}{\cap}} (A_{v_i} \cup A_{\sigma_i} \cup A_{\eta_i})$. Thus the set

$\{x(t); t(b) = 1\} \subseteq \cap (A_{v_i} \cup A_{\sigma_i} \cup A_{\eta_i}) = C$. But if $x \in C$ we can define a truth function by $t(a_v) = 1$ iff $x \in A_v$, and as in lemma (2-8), t satifies $t(b) = 1$. Hence $\{x(t); t(b) = 1\} = C$. The map $t \to x(t)$ is one-one therefore $\mu(C) = J\ 3^{-m}$, where J is the number of distinct truth functions which satisfy $t(b) = 1$.

Let p^0, p^1, be p^J for $J = 0$, $J = 1$ respectively, we can now prove

Theorem (2-10): b is a logical falsity if and only if $p^1 \notin c(n,S)$

Proof: First note that we always have $p^0 \in c(n,S)$. To see this take (X, Σ, μ), $A_1 \dots A_k$, $B_1 \dots B_m$ as in lemma (2-9) but this time choose $C = \phi$. Now suppose $p^1 \in c(n,S)$ then by lemma (2-8) there exist a truth function t such that $t(b) = 1$ and b is not a logical falsity.

Suppose that b is not a logical falsity, then there exist t such that $t(b) = 1$.

Let $J \geq 1$ be the number of such truth functions, then $p^J \in c(n,S)$ by lemma (2–9).

But $p^1 = (1 - \frac{1}{J})p^0 + \frac{1}{J}p^J$ and $p^0 \in c(n,S)$, $c(n,S)$ is convex and therefore

$p^1 \in c(n,S)$.

This leads straightforwardly to the conclusion:

Corrolary (2–11): If condition SIMP$_3$ is satisfied then there is a positive solution to problem (2–7) that is, if there is a polynomial time algorithm which decides whether a given p is an element of c(n,S) then there is a polynomial time algorithm to decide whether a given simple b is a logical falsity.

Proof: Given a simple porposition b we construct the vector p^1. This takes a number of steps which is a polynomial in k,m (remember that the codes for the numbers $3^{-l(v)}$, 3^{-m} have lengths $l(v)$, m respectively). Then we use the polynomial time algorithm to decide whether $p^1 \in c(n,S)$ or not. Now b is a logical falsity iff $p^1 \notin c(n,S)$.

This means that SIMP$_3$ is probably false, or at least that deciding membership in c(n,S) is "as hard as" the tautology problem. But SIMP$_2$ is also probably false to see that note that b is a logical falsity iff ~b is a tautology. Now consider the following:

<u>Problem (2-12)</u>: <u>Does there exist a fixed polynomial R(x) such that for every</u> <u>tautology of the form ~b, b simple, there exists a proof whose length does not</u> <u>exceed R(|b|)?</u>

Problem (2-12) is different from problem (2-7). There may *exist* a short proof for every tautology of the form ~b but we may not be able to *find* that proof in polynomial time.

Clearly an affirmative answer to problem (2-6) entails an affirmative answer to problem (2-12) but not vice versa. Anyway, the answer to problem (2-12) is also unknown, and most researchers believe the answer to be negative. Now if $SIMP_2$ is valid then problem (2-12) has positive answer. For suppose ~b is a tautology then b is a logical falsity and this happens iff $p^1 \notin c(n,S)$. But $p^1 \notin c(n,S)$ iff at least one of the facet inequalities of $c(n,S)$ that is, one of the inequalities in $I(n,S)$, is violated. Since the production of each inequality in $I(n,S)$ takes polynomial time (by $SIMP_2$) there exists a proof of ~b, which consists of a demonstration that one inequality of $I(n,S)$ is violated by p^1. The length of such a proof is the production time of the inequality and the time it takes to demonstrate that the inequality is violated. Again one must remeber that *finding* the relevant inequality may very well take exponential time but this need not concern us. The only relevant feature for Problem (2-12) is that such an inequality exists and its production (by itself) requires only polynomial time.

To put matters in simple language: What all this deomonstrates is that producing all facet inequalities of $c(n,S)$ will probably require exponential time.

More precisely the task seems to be as complex as the tautology problem for the propositional caculus.

2.8 Correlations and the Entropy Principle

Every correlation vector $p \in c(n,S)$ can be represented as a convex combination of the 2^n vertices u^ε, $\varepsilon \in \{0,1\}^n$ that is, there are numbers $\lambda(\varepsilon) \geq 0$ such that:

$$\sum_{\varepsilon \in \{0,1\}^n} \lambda(\varepsilon) = 1 \ , \quad \sum_{\varepsilon \in \{0,1\}^n} \lambda(\varepsilon)\varepsilon_i = p_i \quad 1 \leq i \leq n, \tag{2-23}$$

$$\sum_{\varepsilon \in \{0,1\}^n} \lambda(\varepsilon)\varepsilon_i\varepsilon_j = p_{ij} \ , \quad \{i,j\} \in S$$

Usually there exist more than one choice of coefficients $\lambda(\varepsilon) \geq 0$ which satisfy equations (2-23) (except for the case $n = 2$ and $S = \{\{1,2\}\}$ where the $\lambda(\varepsilon)$ are uniquely fixed by the values of p_1, p_2, p_{12}).

The reason for the existence of multitude of representations is clear from a probablistic standpoint. The numbers p_i, p_{ij} represent the probabilities $\mu(A_i)$, $\mu(A_i \cap A_j)$. Usually these values do not determine uniquely the value of $\lambda(\varepsilon) = \mu(A_1^{\varepsilon_1} \cap A_2^{\varepsilon_2} \cap ... \cap A_n^{\varepsilon_n})$ (in the notations of theorem (2-3)).

By Carathéodory's theorem we can always choose $\lambda(\varepsilon)$ in such a way that at most dim $R(n,S) + 1 = n + |S| + 1$ of them are non zero. Again such a choice, too is usually not unique

Given $p \in c(n,S)$ how are we to fix the values of $\lambda(\varepsilon)$? There is no general way to do this but in some cases the following rule is justified:

Maximum Entropy Principle: given $p \subseteq c(n,S)$, find 2^n numbers $\lambda(\varepsilon) \geq 0$, $\varepsilon \in \{0,1\}^n$ such that the functional

$$H = - \sum_{\varepsilon \in \{0,1\}^n} \lambda(\varepsilon) \log \lambda(\varepsilon) \qquad (2\text{--}24)$$

is maximal under the constraints:

$$\sum_{\varepsilon \in \{0,1\}^n} \lambda(\varepsilon) = 1, \quad \sum_{\varepsilon \in \{0,1\}^n} \lambda(\varepsilon)\varepsilon_i = p_i \quad i = 1,2,...,n$$

and

$$\sum_{\varepsilon \in \{0,1\}^n} \lambda(\varepsilon)\varepsilon_i\varepsilon_j = p_{ij} \quad \{i,j\} \in S$$

In some cases a unique solution to the above convex program exists. When is it justified to use this rule? The concept of "entropy" like "propability" has various interpretations. Under the "objective" view of probability as a representation of frequencies or proportions of properties, the entropy is a measure of the amount of mixing, or interdependence of these properties. In that view the application of the maximum entropy principle is justified when we have good reasons to assume that, apart from the correlations p_{ij}, $\{i,j\} \in S$, the system is maximally mixed, and the properites maximally independent.

Under the "subjective" interpretation of probability the entropy (or rather negentropy) is a measure of information. Thus the application of the maximum entropy principle is justified when we have good reasons to believe that the correlations p_{ij}, $\{i,j\} \in S$ represent all the information available about the system.

Consider as an example the case $n = 3$ and $S = \{\{1,2\}, \{1,3\}, \{2,3\}\}$. In this case $c(n,S)$ is the Bell–Wigner polytope. We proved in theorem (2–4) that given $p = (p_1, p_2, p_3, p_{12}, p_{13}, p_{23}) \in c(n,S)$ the values of $\lambda(\varepsilon)$, $\varepsilon \in \{0,1\}^n$ depend on one parameter η which satisfies inequalities (2–13) namely:

$$\eta \leq \min\{p_{12}, p_{13}, p_{23}, 1 - (p_1 + p_2 + p_3 - p_{12} - p_{13} - p_{23})\}$$

$$\eta \geq \max\{0, -p_1 + p_{12} + p_{13}, -p_2 + p_{12} + p_{23}, -p_3 + p_{13} + p_{23}\}$$

Once such η is fixed then the $\lambda(\varepsilon)$ are given by (2–14). If we substitute the values of $\lambda(\varepsilon)$ from (2–14) into $H(\eta) = -\sum_{\varepsilon \in \{0,1\}^3} \lambda(\varepsilon) \log \lambda(\varepsilon)$ and set $\dfrac{dH}{d\eta} = 0$ we obtain a 4th degree equation in η:

$$[1 - (p_1 + p_2 + p_3 - p_{12} - p_{13} - p_{23}) - \eta][p_{12} - \eta][p_{13} - \eta][p_{23} - \eta] =$$
$$= \eta[\eta + (p_1 - p_{12} - p_{13})][\eta + (p_2 - p_{12} - p_{23})][\eta + (p_3 - p_{13} - p_{23})].$$

And a solution to this equation, which satisfies the constraints on η in (2–13) fixes the values of $\lambda(\varepsilon)$.

2.9 Notes and Remarks

The theory of convex–sets is developed in Rockafeller (1970) and convex polytopes in Yemelichev, Kovalev and Kravtsov (1984).

Constraints on the possible range of values of correlations, in the form of inequalities, have been investigated for many years. Inequalities (2–19) in a more general form were obtained by Bonferroni (1936a, 1936b). These and many other related results are summarized in a monograph by Frechet (1941). Physicists have developed an interest in that subject especially since the work of Bell (1964) who demonstrated that quantum frequencies violate certain such inequalities. The physical aspects of Bell's work and those who further developed the subject will be taken up in subsequent chapters. Bell's work was simplified by Wigner (1969) and Belifante (1973). These authors have not used inequalities (2–12) explicitely, but the name "Bell–Wigner" polytope is justified since their work deals essentially with three events and their joints. Inequalities equivalent to (2–15) where derived by Clauser, Horne, Shimony and Holt (1969). There the inequalities are expressed in terms of expectation values. The verison which appears here is according to Clauser and Horne (1974).

It should be noted that in the work of the physicists mentioned so far, and in the papers of many subsequent authors, the physical aspects of the problems are intermingled with the purely mathmatical character of the derivation of the inequalities. This is a source of prevailing confusion, as if Bell type inequalities have, in themselves something to do with physics. But they do not. I hope the reader is already convinced that these inequalities follow directly from the theory of probability or, if you like, from propositional logic. It is only their violation by quantum frequencies which makes them important for the foundations of physics.

Fine (1982a) proved theorems (2–4) and (2–5). The proof given above is slightly different since it is based on geometry rather than probability. Fine's paper should also be noted as the first to make the purely probablistic character of the constraints evident.

Mermin and his students have generalized the iequalities to higher dimensional cases: Mermin and Shwarz (1982) and Garg and Mermin (1982a, 1983) and in particular (1984). In this last paper the conection between Bell type inequalities and linear programming is made explicit for the first time.

Correlation polytoes were introduced in Pitowsky (1986). There the connection between these polytopes and propositional logic is indicated and theorem (2–3) is proved. The symmetries associated with c(n,S) and the inequalities (2–20) appear in Pitowsky (1988).

The nature of the set of *all* inequalities for a particular fixed dimension was a center of a short exchange between Garg and Mermin (1982b) and Fine (1982b). The results of Pitowsky (1988) which are partially reproduced in section 2.7 "settle" the dispute by demonstrating that deriving all inequalities in the general case is an extremely complex task, probably too complex to be carried out in practice.

The concepts of computational complexity relevant to a deeper understanding of section 2.7 are given in the very readable textbook by Garey and Johnson

(1979). Problem (2-6) is the notorious NP $\overset{?}{=}$ P question. The (polynomial time) equivalence of Problem (2-6) and Problem (2-7) is proved in Shaeffer (1978). Problem (2-12) is the NP $\overset{?}{=}$ coNP question. Theorem (2-10) and the lemmas preceeding it are in Pitowsky (1988). In the language of computational complexity what has been demonstrated is that $SIMP_3 \to P = NP$ and $SIMP_2 \to NP = coNP$. The fact that $SIMP_1$ is valid is due to a general result of Karp and Papadimitriou (1980).

My motivation for developing the theory of correlation polytopes is by and large the foundations of quantum theory. We shall make extensive use of the results of this section in subsequent sections. It should be noted however that the theory of correlation polytopes has direct applications to other fields of studies. In physics it is tightly related to combinatorial problems associated with the Ising spin model. It may also have some important consequence in the theory of computation as well. All these are indicated in Pitowsky (1988).

The concept of correlation polytope may be easily extended to include not just pair correlations but also tripple correlations $p_{ijk} = \mu(A_i \cap A_j \cap A_k)$, quadruple correlations p_{ijkl} and so forth. With obvious extensions of our defintions all the general results remain valid.

3. Quantum Correlations

3.1. Introduction

Quantum mechanics provides us with a machinary for calculating probabilities of microscopic events. The rigorous account of that machinary was developed by von Neumann in his classic "The Mathematical Foundations of Quantum Mechanics." We shall adopt his approach here.

One striking feature of the Hilbert space formalism lies in the fact that, prima-facie, it does not resemble at all the standard classical approaches to probability. Even though one purpose of quantum theory is the calculation of probabilities, frequencies, cross-sections, and so forth, these calculations make no explicit use of terms like "probability spaces", "random variables", "independent events" and other such concepts, which populate standard texts of probability theory. A-priori there are two possibilities: Either the Hilbert space formalism is just a sophisticated guise for standard probablistic reasoning or else it is something entirely different. Of course the second possibility is the correct one. The best way to understand the difference between quantum probability and classical probability is in terms of correlations. It turns out that the Hilbert space formalism relaxes considerably the constraints on correlations which exist in the classical theory. Often quantum frequencies violate the facet inequalities of $c(n,S)$. These are typically cases in which physicists talk about "interference".

In this chapter I shall introduce, in a general way, the probablistic machinary of quantum mechanics (Sections 3.2, 3.3) then identify the range of possible values of quantum correlations, again, in terms of convex polytopes (Section 3.5). Consequently I shall analyze some special cases, notably cases in which there is an apparent violation of locality (Sections 3.6, 3.7, 3.8, 3.9). Finally, I shall briefly describe some possible interpretations of these results.

3.2 Probability in Quantum Mechanics

In quantum theory we associate with every physical system a complex Hilbert space H. At every given moment the system is said to be in a given state. The concept of state is given by

Definition (3–1) A bounded linear operator W on H is called a state if (i) W is Hermitian, $W^t = W$, (ii) W is semi–definite that is, $<\phi, W\phi> \geq 0$ for all $\phi \in H$, (iii) W is trace class and $tr(W) = 1$

For a given fixed unit vector $\phi \in H$ let E_ϕ denote the (orthogonal) projection operator on ϕ. If $W = E_\phi$ for some unit vector ϕ then the state W is said to be *pure*, otherwise it is called a *mixture*. The name is justified by the following

Lemma (3-1): W is a state on H if and only if there are unit vectors

$\phi_1, \phi_2, ..., \phi_n ...$ and non negative numbers $\lambda_1, \lambda_2, ..., \lambda_{n_1} ...$ such that $\sum_{n=1}^{\infty} \lambda_n = 1$

and $W = \sum_{n=1}^{\infty} \lambda_n E_{\phi_n}$.

Proof: If $W = \sum_{n=1}^{\infty} \lambda_n E_{\phi_n}$ where $\lambda_n \geq 0$ and $\sum_{n=1}^{\infty} \lambda_n = 1$ and ϕ_n are

normalized then clearly $W^t = W$ since every projection is Hermitian, also

$<\phi, W\phi> = \sum_{n=1}^{\infty} \lambda_n |<\phi, \phi_n>|^2 \geq 0$, and $tr(W) = \sum_{n=1}^{\infty} \lambda_n tr(E_{\phi_n}) = \sum_{n=1}^{\infty} \lambda_n = 1$.

Conversely, suppose W is a state then the spectrum of W consists only of

egenvalues. (The proof of this fact is in Von Neumann, 1955, p. 198, footnote 115,

which I shall not reproduce here since we are mainly concerned with finite

dimensional cases anyway.)

Let $\lambda_1, \lambda_2, ..., \lambda_n ...$ be the eigenvalues (which may not be all distinct in case

of degeneracy) which corresponds to the orthonormal eigenvalues $\phi_1, \phi_2, ..., \phi_n ...$

respectively. Then clearly $W = \sum_{n=1}^{\infty} \lambda_n E_{\phi_n}$, $\lambda_n \geq 0$ since W is Hermitian

semi-definite, and $\sum_{n=1}^{\infty} \lambda_n = tr(W) = 1$.

Every measurable magnitude or observable, associated with a physical system

(such as energy, momentum, angular momentum, spin and so forth) is represented

in quantum mechanics by an Hermitian operator on H. In the following I shall

somewhat loosely identify the observable with the operator which represents it.

Thus if A is the operator I shall speak about "the value of A" in a given state

"the expectation of A" in a given state and so forth.

The operational rule underlying quantum probability calculations is the following:

Given a system in a state W and an observable A the expectation of A in the state W is:

$$<A>_w = tr(WA) \tag{3-1}$$

In particular if $W = E_\phi$, where ϕ is a normalized vector then the expectation of A in this state, denoted by $<A>_\phi$, is given by $<A>_\phi = <\phi, A\phi>$.

Every physical observable is associated with a Hermitian operator. What about the opposite claim? Does every Hermitian operator correspond to an observable? Clearly this is not the case, superselection rules testify to that. But even when we ignore the usual superselection rules the question does not disapear. Thus, it has been questioned whether all Hermitian operators in the position-momentum algebra represent observables (Wigner, 1976). Fortunately we will not have to deal with this problem. The difficulties associated with quantum probability arise in contexts where no problem of measurability exist. Most of our examples will be associated with spin and angular momentum algebras where all measurements are well defined (though perhaps, difficult to perform).

Let W be a state and A an observable. Following the example of classical probability theory we define the standard deviation of A in the state W by:

$$\Delta A = [<A^2>_w - <A>_w^2]^{1/2} \tag{3-2}$$

with this definition we can prove:

Theorem (3-2) (The Uncertainty Relations): Let A, B be two observables then in the state $W = E_{\phi_1}$ ϕ normalized, we have

$$\Delta A \cdot \Delta B \geq \frac{1}{2}|<\phi, \ [A,B]\phi>| \qquad (3-3)$$

where $[A,B] = AB - BA$ is the commutator

Proof: $W = E_\phi$ is fixed, put $\hat{A} = A - <A>_\phi$, $\hat{B} = B - <B>_\phi$, then $[\hat{A},\hat{B}]$ $= [A,B]$ and $\Delta\hat{A} = \Delta A = <\hat{A}^2>$, $\Delta\hat{B} = \Delta B = <\hat{B}^2>$. By Shwarz inequality

$$(\Delta A)^2(\Delta B)^2 = <\phi, \ \hat{A}^2\phi><\phi, \ \hat{B}^2\phi> = <\hat{A}\phi, \ \hat{A}\phi><\hat{B}\phi, \ \hat{B}\phi> \geq \qquad (3-4)$$

$$\geq |<\hat{A}\phi, \ \hat{B}\phi>|^2 = |<\phi, \ \hat{A}\hat{B}\phi>|^2$$

But

$$\hat{A}\hat{B} = \frac{\hat{A}\hat{B} + \hat{B}\hat{A}}{2} + \frac{\hat{A}\hat{B} - \hat{B}\hat{A}}{2} = \frac{\hat{A}\hat{B} + \hat{B}\hat{A}}{2} + \frac{1}{2}[A,B]$$

Hence

$$<\phi,\hat{A}\hat{B}\phi> = <\phi, \ \frac{\hat{A}\hat{B} + \hat{B}\hat{A}}{2}\phi> + \frac{1}{2} <\phi,[A,B]\phi> \qquad (3-5)$$

The number $<\phi, \dfrac{\hat{A}\hat{B} + \hat{B}\hat{A}}{2}\phi>$ is real since $\dfrac{\hat{A}\hat{B} + \hat{B}\hat{A}}{2}$ is Hermitian. Similarily

$<\phi, [A,B]\phi>$ is pure imaginary since $[A,B]$ is anti–Hermitan. $[A,B]^t = (AB - BA)^t$

$= -[A,B]$ and thus $<\phi, [A,B]\phi> = -<[A,B]\phi,\phi>$

$= -<\phi, [A,B]\phi>^*$ which means that $\mathrm{Re}<\phi,[A,B]\phi> = 0$. Thus from (3–5) we have

$$|<\phi, \hat{A}\hat{B}\phi>|^2 = |<\phi, \dfrac{\hat{A}\hat{B} + \hat{B}\hat{A}}{2}\phi>|^2 + \dfrac{1}{4}|<\phi, [A,B]\phi>|^2 \geq \dfrac{1}{4}|<\phi, [A,B]\phi>|^2.$$

Substituting in (3–4) we conclude:

$$\Delta A \Delta B \geq \dfrac{1}{2}|<\phi, [A,B]\phi>|$$

3.3 The Lattice of Closed Subspaces of a Hilbert Space

Among the observables of quantum mechanics the simplest are those which correspond to *properties* which the system may have (in which case the value of the observable is 1) or lack (in which case the value of the observable is zero). Such observables are represented by (orthogonal) projections on closed subspaces of H.

Let A be any observable. In the general case we can represent A by its spectral resolution $A = \int_{\sigma(A)} xE(dx)$, where $E(dx)$ is the spectral measure corresponding to A, and $\sigma(A)$ the spectrum of A. We shall be concerned mainly with operators which have point spectrum, in fact with finite dimensional cases, so let us concentrate on the simple case where A can be represented as $A = \sum_{n=1}^{\infty} x_n E_n$ where the x_n's are the eigenvalues of A and E_n is the projection on the space spanned by the eigenvectors corresponding to the eigenvalue x_n. Thus the projection E_n corresponds to the property: "The observable A has value x_n". Let W be the state of the system and suppose $W = E_\phi$ for some normalized ϕ, then the system has the property associated with E_n if and only if

$tr(WE_n) = tr(E_\phi E_n) = 1$. This happens if and only if $E_n\phi = \phi$ that is, ϕ itself is an eigenvector of A corresponding to the eigenvalue x_n.

Since we adopt the idealization that every Hermitian operator represents an observable we conclude that every (closed) projection corresponds to a property. Again I shall use a somewhat loose formulation and speak of a system "having property E".

Among the projections (remember we deal only with orthogonal projections on closed subspaces) there are certain relations and operations which correspond to logical relations and operations among properties.

(a) Given two projections E_1, E_2 we shall denote $E_1 \leq E_2$ in case $E_1(H) \subseteq E_2(H)$. If $E_1 \leq E_2$ then every eigenstate of E_1 is necessarily an eigenstate of E_2. Thus $E_1 \leq E_2$ corresponds to "E_1 entails E_2". We have $E_1 \leq E_2$ if and only if $tr(WE_1) = 1$ entails $tr(WE_2) = 1$ for all states W.

(b) Let E_1, E_2 be two, not necessarily commuting, projections. Let $E_1 \wedge E_2$ denote the projection onto the subspace $E_1(H) \cap E_2(H)$. Clearly ϕ is an eigenstate of $E_1 \wedge E_2$ if and only if ϕ is an eignstate of both E_1 and E_2. Therefore, $E_1 \wedge E_2$

is the property corresponding to "E_1 and E_2". We have $\text{tr}[W(E_1 \wedge E_2)] = 1$ if and only if $\text{tr}(WE_1) = \text{tr}(WE_2) = 1$, for all states W.

These essentially are the *logical* relations among properties which are derived in Hilbert space. As always the relation "entails" corresponds to the set–theoretic inclusion (among subspaces) and "and" corresponds to the set theoretic intersection (of subspaces). What about "not E" and "E_1 or E_2"? If E is a projection the set theoretic complement of E(H) is *not* a subspace. the "natural" candidate to represent "not E" is the projection $E^{\perp}$ on the orthogonal complement of E(H) and indeed we have $\text{tr}(WE) = 0$ iff $\text{tr}(WE^{\perp}) = 1$, for all states W. But since $E^{\perp}(H)$ is not the set theoretic complement of E(H), the identification of $E^{\perp}$ with "not E" is a problematic move. Similarly we can define $E_1 \vee E_2$ to be the projection on the closed subspace *spanned by* $E_1(H) \cup E_2(H)$. With this definition $E_1 \vee E_2 = (E_1^{\perp} \wedge E_2^{\perp})^{\perp}$ that is, De Morgan rule is satisfied. But again, since $(E_1 \vee E_2)(H)$ is not the set theoretic union of $E_1(H)$ and $E_2(H)$ the identification of $E_1 \vee E_2$ with "E_1 or E_2" is a problematic matter.

Birkhoff and von Neumann (1942) have made the identification of $E^{\perp}$ with the property "not E", and the identification of $E_1 \vee E_2$ with the property "E_1 or E_2". The consequence is a violation of some of the rules of classical logic. According to Birkhoff and von Newman, this is an indication that the laws of classical logic may not be suitable for microphysical discourse, and should therefore be substituted by the new rules of "quantum logic."

This type of argument suffers from the disadvantage of being totally dependent on the details of the Hilbert space formalism (or some of its extentions such as C*-algebras). For that reason Birkhoff and Von Newmann themselves called their approach "heuristic."

In chapter 4 I shall develop the details of quantum logic from a completely different perspective. The approach to be adopted there is directly connected to the numerical values of quantum correlations, and therefore allows for a detailed examination of the *physical* content of quantum logic.

Anyway, at this stage I should stress that only the relation "≤" and the operation "∧" among projections are taken to have logical meaning; and indeed they have *classical* logical meaning. The operations ⊥ and "∨" are to be understood as geometric operations, divorced of logical meaning. Given a Hilbert space H let L(´l) be the set of all (orthogonal) projections on (closed) subspaces of H. L(H) together with the relation ≤ the operations, ∧, ⊥, ∨, the null projection 0, and the unit projection I is an *orthocomplemented lattice* that is, the following axioms are satisfied for all E, E', E" ∈ L(H):

$$0 \neq I \quad 0 \leq I \qquad\qquad (3\text{--}6)$$
$$E \leq E$$
if E ≤ E' and E' ≤ E" then E ≤ E"
if E ≤ E' and E' ≤ E then E = E'
$$0 \leq E \leq I$$

$$E \vee E = E, \ E \vee E' = E' \vee E, \ (E \vee E') \vee E" = E \vee (E' \vee E")$$
$$E \vee 0 = E, \ E \vee I = I$$
$$E \vee E' = E \text{ iff } E' \leq E$$

$$E \wedge E = E, \ E \wedge E' = E' \wedge E, \ (E \wedge E') \wedge E" = E \wedge (E' \wedge E")$$
$$E \wedge 0 = 0, \ E \wedge I = E$$
$$E \wedge E' = E \text{ iff } E \leq E'$$

$$E \wedge E^{\perp} = 0 \quad E \vee E^{\perp} = I$$
$$(E^{\perp})^{\perp} = E$$
$$(E^{\perp} \vee E'^{\perp})^{\perp} = E^{\perp} \wedge E'^{\perp}$$

In addition if H is finite dimensional L(H) is *modular* that is:

If $E \leq E''$ then $E \vee (E' \wedge E'') = (E \vee E') \wedge E''$ (3–7)

(3–7) does not hold if H is not finite dimensional. A subset $T \subseteq L(H)$ is called a sublattice if $0 \in T$ and T is closed under the lattice operations $\vee$, $\wedge$, $\perp$. A sublattice T is boolean if for all E, E', $E'' \in T$

$$E \wedge (E' \vee E'') = (E \wedge E') \vee (E \wedge E'') \qquad\qquad (3\text{–}8)$$

(3–8) is called "the distributivity rule". It is easy to see that if a sublattice T of L(H) is boolean, then all its elements commute and, in this case we have:

$$E \wedge E' = EE' = E'E \text{ and } E \vee E' = E + E' - EE'$$

3.4   A Note on Gleason's Theorem

The distinction between pure states and mixtures is really based on an analogy with classical statistical mechanics. Given a classical system, a state of maximal knowledge regarding the system is represented by a single point in the phase space associated with the system, or alternatively by an atomic (Dirac) measure concentrated on a single point. When no definite precise information about the system is available the states are represented by a probability measure on (a hypersurface in) phase space. By analogy a pure state in quantum theory is represented by a projection on a nuit vector in Hilbert space and mixtures by convex combinations of pure states.

There is one important disanalogy here, surely not every probability measure in phase space is a convex combination of atomic measures. Maybe one should consider more general probability measures on Hilbert space too? It turns out that there are no "more general" measures! Consider the following:

Definition (3–2): Let L(H) be the lattice of closed orthogonal projections on a seperable Hilbert space H. A non negative real function μ on L(H) is called a probability measure if $\mu(I) = 1$ and if whenever $E_1, E_2, ..., E_n ...$ are pairwise orthogonal projections, and E the projection on the closed subspace spanned by

$$\bigcup_{i=1}^{\infty} E_i(H) \text{ then } \mu(E) = \sum_{n=1}^{\infty} \mu(E_n)$$

In other words, μ is a probability measure on L(H) iff its restriction to every Boolean sublattice of L(H) is a probability measure in the usual sense. The following remarkable theorem is due to Gleason (1957):

Theorem (3-3): If H is a real or complex Hilbert space of dimension greater or equals 3, and μ a probability measure on L(H), then there exists a state W on H such that $\mu(E) = tr(WE)$ for all $E \in L(H)$.

Our considerations will not depend on Gleason's theorem and the details of its proof. It is important to note that as a result of Gleason's theorem, our discussion will not be limited in scope. We shall identify the constraints on quantum correlations in terms of states W, and Gleason's result assures us that these essentially are all theoretically possible states. More on Gleason's theorem in the "Notes and Remarks" section.

3.5 Quantum Correlation Polytopes

As in our discussion of classical correlation polytopes in the previous chapter, let $n \geq 2$ be a natural number S a set of pairs of numbers:
$S \subset \{\{i,j\} \mid 1 \leq i < j \leq n\}$, $R(n,S)$ the real space of all functions
f: $\{1, 2, \ldots n\} \cup S \to R$. Keeping the same notations, we define by anology with the classical case, expressed in theorem (2-3):

Definition (3-3): Let $p = (p_{11}, \ldots, p_{n1} \ldots p_{ij} \ldots) \in R(n,S)$. We shall say that p has a quantum mechanical representation if there exists a Hilbert space H, a state W

on H, and (not necessarily commuting, not necessarily distinct) projections
$E_1, ..., E_n \in L(H)$ such that:

$$p_i = tr(WE_i) \quad 1 \leq i \leq n , \quad p_{ij} = tr[W(E_i \wedge E_j)] \quad \{i,j\} \in S \quad (3-9)$$

Definition (3-3) is indeed an analogue of the classical case expressed in theorem
(2-3). We have seen that $E_1 \wedge E_2$ does represent the (classical) conjunction of
properties "E_1 and E_2". Therefore if p has a quantum mechanical representation,
p_{ij} is indeed the probability that the system has the conjunctive property
"E_i and E_j".

Let q(n,S) be the set of all vectors $p \in R(n,S)$ which have quantum
mechanical representation. Let l(n,S) be the set of all vectors
$p = (p_1 ... p_n ... p_{ij} ...) \in R(n,S)$ which satisfy:

$$0 \leq p_i \leq 1 \quad 1 \leq i \leq n$$
$$0 \leq p_{ij} \leq \min\{p_i, p_j\} \quad \{i,j\} \in S \quad (3-10)$$

Then l(n,S) is a closed convex polytope (it is a bounded set of vectors which
satisfy linear constraints). Our aim is to prove that the following relations obtain
between c(n,S), q(n,S), l(n,S):

65

Theorem (3-4):

    (a) $c(n,S) \subseteq q(n,S) \subseteq l(n,S)$

    (b) $q(n,S)$ is convex (but not closed);

    (c) $q(n,S)$ contains the interior of $l(n,S)$.

Theorem (3-4) (c) states that $q(n,S)$ is essentially $l(n,S)$ save for some points on the boundary of $l(n,S)$. (The points of $l(n,S)\backslash q(n,S)$ will be identified later.) Theorem (3-4) (a) states that every classical $p \in c(n,S)$ has a quantum mechanical representation, so that the Hilbert space formalism is more flexible, or less contrained than the classical formalism. The proof of theorem (3-4) is quite involved and will follow some lemmas.

    Lemma (3-5): $l(n,S)$ is a closed convex polytope whose vertices are all the vectors $u = (u_1, ..., u_n ... u_{ij} ...) \in R(n,S)$ such that $u_i, u_{ij} \in \{0,1\}$ for $1 \le i \le n$, $\{i,j\} \in S$, and such that $u_{ij} \le \min\{u_i, u_j\}$ for all $\{i,j\} \in S$.

    Proof: It is clear that $l(n,S)$ is closed and convex. Suppose that $u \in R(n,S)$ satisfy $u_i, u_{ij} \in \{0,1\}$ and $u_{ij} \le \min\{u_i, u_j\}$ then clearly $u \in l(n,S)$. Also u is an extreme point of $l(n,S)$ for if $u = \lambda p + (1 - \lambda)p'$ for $p, p' \in l(n,S)$ and $0 < \lambda < 1$ we have $u_i = \lambda p_i + (1 - \lambda)p_i'$ $1 \le i \le n$ and $u_{ij} = \lambda p_{ij} + (1 - \lambda)p_{ij}'$ $\{i,j\} \in S$. Since $0 \le p_i \le 1$ and $0 \le p_{ij} \le 1$ and since

u_i, u_{ij} $\in$ {0,1} it follows that $p_i = p'_i = u_i$ and $p_{ij} = p'_{ij} = u_{ij}$ for $1 \leq i \leq$ n and

{i,j} $\in$ S. Hence p = p' = u and u is an extreme point of l(n,S).

Conversely suppose that u is an extreme point of l(n,S) we shall show that

u_i, u_{ij} $\in$ {0,1} and $u_{ij} \leq$ min{u_i, u_j}. Consider the function $f(t) = t^2$. It is an

increasing function in the interval [0,1], f(0) = 0 and f(1) = 1. Now define

p $\in$ R(n,S) by $p_i = f(u_i)$ $p_{ij} = f(u_{ij})$ for $1 \leq i \leq$ n and {i,j} $\in$ S. Then from the

above properties of f it follows that p $\in$ l(n,S). Consider $g(t) = 2t - t^2$. It is an

increasing function on [0,1], g(0) = 0, g(1) = 1 therefore if we define

p' $\in$ R(n,S) by $p'_i = g(u_i)$, $p'_{ij} = g(u_{ij})$ we have p' $\in$ l(n,S). But $\frac{1}{2}(f(t) + g(t)) = t$

hence u = $\frac{1}{2}$p + $\frac{1}{2}$p'. Since u is an extreme point of l(n,S) it follows that

u = p = p', hence $u_i^2 = u_i$ $u_{ij}^2 = u_{ij}$ and therefore u_i, u_{ij} $\in$ {0,1} for

$1 \leq i \leq$ n and {i,j} $\in$ S also $u_{ij} \leq$ min{u_i, u_j}, {i,j} $\in$ S since u $\in$ l(n,S). Every

convex compact subset in Euclidean space is the convex hull of its extreme points

and therefore l(n,S) is a polytope whose vertices are just the vectors u with the

above properties.

Examples: (a) Let n = 2 and S = {{1,2}} then l(n,S) has five vertices

(u_1, u_2, u_{12}) namely: (0,0,0), (1,0,0), (0,1,0), (1,1,1) and (1,1,0). Note that the first

four vertices are the truth values for two propositions and their conjuction. In

other words, they are the vertices of c(2,S). We add to them a fifth vertex

(1,1,0).

(b) Let n = 3 and S = {{1,2}, {1,3}, {2,3}} then l(3,S) has 18 vertices

$(u_1, u_2, u_3, u_{12}, u_{13}, u_{23})$. The eight vertices of c(3,S): (0,0,0,0,0,0), (1,0,0,0,0,0), (0,1,0,0,0,0), (0,0,1,0,0,0), (1,1,0,1,0,0), (1,0,1,0,1,0), (0,1,1,0,0,1), (1,1,1,1,1,1) and additional 10 vertices: (1,1,0,0,0,0), (1,0,1,0,0,0), (0,1,1,0,0,0), (1,1,1,1,1,0), (1,1,1,1,0,1), (1,1,1,0,1,1), (1,1,1,1,0,0), (1,1,1,0,1,0), (1,1,1,0,0,1), (1,1,0,0,0,0).

In general the number of vertices of l(n,S) is calculated in the following way: Given a set $\alpha \subset \{1,2 \ldots n\}$ let $\alpha * \alpha$ denote the set of pairs $\alpha * \alpha = \{\{i,j\}, i,j \in \alpha, i < j\}$. As usual $|\alpha * \alpha \cap S|$ denotes the cardinality of $\alpha * \alpha \cap S$. The number of vertices of l(n,S) is $1 + n + \Sigma \, 2^{|\alpha * \alpha \cap S|}$ where the sum is taken over all subsets $\alpha \subset \{1,2, \ldots n\}$ such that $|\alpha| \geq 2$. The proof is left as an exercise to the reader. In case n = 4 and S = {{1,3}, {1,4}, {2,3}, {2,4}}, l(n,S) has 47 vertices.

We shall prove now a geometrical lemma about Hilbert spaces. Let H, H' be two Hilbert spaces. We shall denote by $H \oplus H'$ the direct sum of H and H' that is, the set of pairs (ϕ, ϕ') where $\phi \in H$ and $\phi' \in H'$, with the vector addition $(\phi_1, \phi_1') + (\phi_2, \phi_2') = (\phi_1 + \phi_2, \phi_1' + \phi_2')$ salar product $\lambda(\phi, \phi') = (\lambda\phi, \lambda\phi')$ and inner multiplication $<(\phi_1, \phi_1'), (\phi_2, \phi_2')> = <\phi_1, \phi_2> + <\phi_1', \phi_2'>$. If E is a projection in H and E' a projection in H' then $E \oplus E'$ denotes the projection on the closed subspace $E(H) \oplus E'(H')$ of $H \oplus H'$.

Similarily let $H \otimes H'$ be the tensor product of H, H' and E, E' projections in H, H' respectively. Let $E \otimes E'$ denote the projection on the subspace $E(H) \otimes E'(H')$ of $H \otimes H'$. We shall prove:

Lemma (3-6): Let H, H' be Hilbert spaces E_1, E_2 projections in H and E_1', E_2' prjections in H' then:

(a) $(E_1 \oplus E_1') \wedge (E_2 \oplus E_2') = (E_1 \wedge E_2) \oplus (E_1' \wedge E_2')$

(b) $(E_1 \otimes E_1') \wedge (E_2 \otimes E_2') = (E_1 \wedge E_2) \otimes (E_1' \wedge E_2')$

Proof: Since we are interested only in finite dimesnional cases we shall assume that H, H' are finite dimensional (though the lemma is true in general)

(a) It is clear that $(E_1 \wedge E_2) \oplus (E_1' \wedge E_2') \leq (E_1 \oplus E_1') \wedge (E_2 \oplus E_2')$ (since $(E_1 \wedge E_2) \oplus (E_1' \wedge E_2') \leq (E_1 \oplus E_1')$ and $(E_1 \wedge E_2) \oplus (E_1' \wedge E_2') \leq (E_2 \oplus E_2')$).

Now suppose $\phi \in (E_1(H) \oplus E_1'(H')) \cap (E_2(H) \oplus E_2'(H'))$. In particular $\phi \in E_1(H) \oplus E_1'(H')$ hence $\phi = (\psi, \psi')$ where $\psi \in E_1(H)$, $\psi' \in E_1'(H')$ but $(E_2 \oplus E_2')\phi = \phi$ so that $\phi = (E_2\psi, E_2'\psi') = (\psi, \psi')$ hence $\|\phi\|^2 = \|\psi\|^2 + \|\psi'\|^2 = \|E_2\psi\|^2 + \|E_2'\psi'\|^2$. But $\|\psi\| \geq \|E_2\psi\|$, $\|\psi'\| \geq \|E_2\psi'\|$ and therefore $\|E_2\psi\| = \|\psi\|$ which entails $E_2\psi = \psi$ and thus $\psi \in E_2(H)$ and therefore $\psi \in E_1(H) \cap E_2(H)$ by the same argument $\psi' \in E_1'(H') \cap E_2'(H')$ and thus $\phi \in (E_1(H) \cap E_2(H)) \oplus (E_1'(H') \cap E_2'(H'))$

(b) First note that $(E_1 \wedge E_2) \otimes (E_1' \wedge E_2') \leq (E_1 \otimes E_1') \wedge (E_2 \otimes E_2')$ by the same argument as in (a). Let I, I' denote the identity operators in H, H' respectively then we have, by definition, for

$E \in L(H)$, $E' \in L(H')$: $E \otimes E' = (E \otimes I')(I \otimes E') = (I \otimes E')(E \otimes I')$.

We shall prove first that for E_1, $E_2 \in L(H)$:

$(E_1 \wedge E_2) \otimes I' = (E_1 \otimes I') \wedge (E_2 \otimes I')$. Suppose

$\phi \in (E_1(H) \otimes H') \cap (E_2(H) \otimes H')$. Let $\xi_1 \ldots \xi_m$ be an orthonormal basis in

$E_1(H)$, $\eta_1 \ldots \eta_n$ an orthonormal basis in $E_2(H)$ and $\psi_1' \ldots \psi_r'$ an orthonormal basis

in H'. Then we can represent

$$\phi = \sum_{j=1}^{r} \sum_{i=1}^{n} a_{ij} \, \xi_i \otimes \psi_j' = \sum_{j=1}^{r} \sum_{i=1}^{m} b_{ij} \eta_i \otimes \psi_j'$$

since $\{\xi_i \otimes \psi_j' \mid 1 \leq i \leq n, \, 1 \leq j \leq r\}$ is an orthonormal basis in $E_1(H) \otimes H'$ and

$\{\eta_i \otimes \psi_j'; \, |1 \leq i \leq m, \, 1 \leq j \leq r\}$ is an orthonormal basis in $E_2(H) \otimes H'$. Let

$1 \leq l \leq r$ be fixed and consider the vector $\phi' = (I \otimes E_{\psi_l}) \phi$ we have:

$$\phi' = \sum_{i=1}^{n} a_{il} \, \xi_i \otimes \psi_l' = \sum_{i=1}^{m} b_{il} \eta_i \otimes \psi_l'$$

hence $\|\phi'\|^2 = \sum_{i=1}^{n} |a_{il}|^2 = \sum_{i=1}^{m} |b_{il}|^2 = \sum_{k=1}^{m} \sum_{i=1}^{n} a_{il} b_{kl}^* \langle \xi_i, \eta_k \rangle$. Now define

$\alpha = \sum_{i=1}^{n} a_{il} \xi_i$, $\beta = \sum_{i=1}^{n} b_{kl} \eta_k$ then $\alpha \in E_1(H)$, $\beta \in E_2(H)$ and we have

$\|\phi'\|^2 = \|\alpha\|^2 = \|\beta\|^2 = \langle \alpha, \beta \rangle$. Therefore $\alpha = \beta$ and in particular

$\alpha = \sum a_{il} \xi_i \in E_1(H) \cap E_2(H)$ therefore

$\alpha \otimes \psi_l' = \sum_{i=1}^{n} a_{il} \xi_i \otimes \psi_l' \in [E_1(H) \cap E_2(H)] \otimes H'$ since this happends for all

$1 \leq l \leq r$ we have also $\phi = \sum_{l=1}^{r} \sum_{i=1}^{n} a_{il} \xi_i \otimes \psi_l' \in [E_1(H) \cap E_2(H)] \otimes H'$. To complete the proof of the theorem suppose that $\phi \in [E_1(H) \otimes E_1'(H')] \cap [E_2(H) \otimes E_2'(H')]$

then in particular $\phi \in [E_1(H) \otimes H'] \cap [E_2(H) \otimes H']$ and therefore

$\phi \in [E_1(H) \cap E_2(H)] \otimes H'$ in the same way $\phi \in [H \otimes E_1'(H')] \cap [H \otimes E'_2(H')]$ and

therefore $\phi \in H \otimes [E_1'(H') \cap E_2'(H')]$. Combining the two results we get $\phi \in [E_1(H)$

$\cap E_2(H)] \otimes [E_1'(H') \cap E_2'(H')]$ and the proof is completed.

We are now ready to prove the main result of this section:

Proof of Theorem (3–4):

(a) We have to show $c(n,S) \subseteq q(n,S) \subseteq l(n,S)$. Let

$p = (p_1 \, \cdots \, p_n \, \cdots \, p_{ij} \, \cdots) \in q(n,S)$ then by definition there are projections

$E_1, \ldots, E_n \in L(H)$ and a state W on H such that $p_i = tr(WE_i)$ $1 \le i \le n$ and

$p_{ij} = tr[W(E_i \wedge E_j)]$, $\{i,j\} \in S$. Hence clearly $0 \le p_{ij} \le min\{p_i, p_j\} \le 1$

and $p \in l(n,S)$ so that $q(n,S) \subseteq l(n,S)$. Let $p \in c(n,S)$ then for each $\varepsilon \in \{0,1\}^n$

there is $\lambda(\varepsilon) \ge 0$ such that $\sum\limits_{\varepsilon \in \{0,1\}^n} \lambda(\varepsilon) = 1$ and $\sum\limits_{\varepsilon \in \{0,1\}^n} \lambda(\varepsilon)\varepsilon_i = p_i$, $1 \le i \le n$, and

$\sum\limits_{\varepsilon \in \{0,1\}^n} \lambda(\varepsilon)\varepsilon_i\varepsilon_j = p_{ij}$ for $\{i,j\} \in S$. Let H be a Hilbert space of dimension 2^n, let

$\{\psi(\varepsilon), \varepsilon \in \{0,1\}^n\}$ be an orthonormal basis in H parametrized by $\varepsilon \in \{0,1\}^n$ in

some arbitrary manner. Let W be the state in H which, relative to the basis

$\{\psi(\varepsilon); \varepsilon \in \{0,1\}^n\}$ is given by the (diagonal) matrix $W(\varepsilon,\varepsilon) = \lambda(\varepsilon)$, $W(\varepsilon,\varepsilon') = 0$ for

$\varepsilon \ne \varepsilon'$. Then clearly $tr(W) = \sum\limits_{\varepsilon \in \{0,1\}^n} \lambda(\varepsilon) = 1$, W is Hermitian and semi–definite.

Let E_i be the projection on the subspace of H spanned by $\{\psi(\varepsilon); \varepsilon_i = 1\}$ then

$E_i \wedge E_j$ is the projection on the subspace spanned by $\{\psi(\varepsilon)|\varepsilon_i = \varepsilon_j = 1\}$ and clearly $\mathrm{tr}(WE_i) = \sum\limits_{\varepsilon\in\{0,1\}^n} \lambda(\varepsilon)\varepsilon_i = p_i$, $1 \le i \le n$ and $\mathrm{tr}[W(E_i \wedge E_j)] = \sum\limits_{\varepsilon\in\{0,1\}^n} \lambda(\varepsilon)\varepsilon_i\varepsilon_j = p_{ij}$

for $\{i,j\} \in S$. Therefore $c(n,S) \subseteq q(n,S)$.

(b) We have to demonstrate that $q(n,S)$ is convex, that is if $p,p' \in q(n,S)$ and $0 < \lambda < 1$ then $\lambda p + (1- \lambda)p' \in q(n,S)$. $p \in q(n,S)$ therefore there is a Hilbert space H, projections E_1, ..., $E_n \in L(H)$ and a state W on H such that $p_i = \mathrm{tr}(WE_i)$ $1 \le i \le n$ and $p_{ij} = \mathrm{tr}[W(E_i \wedge E_j)]$ $\{i,j\} \in S$. Similarily $p' \in q(n,S)$, therefore there are H', E'_1, ..., $E'_n \in L(H')$ and W' on H' such that $\mathrm{tr}(W'E'_i) = p'_i$, $1 \le i \le n$ and $\mathrm{tr}[W'(E'_i \wedge E'_j)] = p'_{ij}$, $\{i,j\} \in S$. Let $\bar{H} = H \oplus H'$ be the direct sum of H and H'. Let $\bar{W}$ be the operator on $\bar{H}$ given by $\bar{W} = (\lambda W) \oplus [(1 - \lambda)W']$ that is, if $\phi \in H$ $\phi = (\psi, \psi')$ for $\psi\in H$, $\psi'\in H'$ then: $\bar{W}\phi = (\lambda W\psi, (1-\lambda)W'\psi')$. Clearly W is Hermitian semi–definite and $\mathrm{tr}(W) = \lambda\mathrm{tr}(W) + (1-\lambda)\mathrm{tr}(W') = 1$. Let $\bar{E}_i = E_i \oplus E'_i$, then by Lemma (3–6) we have: $\bar{E}_i \wedge \bar{E}_j = (E_i \wedge E_j) \oplus (E'_i \wedge E'_j)$ therefore $\mathrm{tr}(\bar{W}\bar{E}_i) = \lambda\mathrm{tr}(WE_i)+ (1-\lambda)\mathrm{tr}(W'E'_i) = \lambda p_i + (1- \lambda)p'_i$ for $1 \le i \le n$, and $\mathrm{tr}[\bar{W}(\bar{E}_i \wedge \bar{E}_j)] = \lambda\mathrm{tr}[W(E_i \wedge E_j)] + (1 - \lambda)\mathrm{tr}[W'(E'_i \wedge E'_j)] = \lambda p_{ij} + (1 - \lambda)p'_{ij}$ for $\{i,j\} \in S$. Hence $\lambda p + (1-\lambda)p' \in q(n,S)$ and $q(n,S)$ is convex.

(c) We have to demonstrate that $q(n,S)$ contains the interior of $l(n,S)$. First note that in order to establish that, it is sufficient to show that $q(n,S)$ is densed in $l(n,S)$. This is the case because for every convex set A in a Euclidean space:

interior (closure (A)) = interior (A). Thus if q(n,S) is densed in l(n,S) we have, since q(n,S) is convex: interior (l(n,S)) $\subseteq$ q(n,S).

Next, in order to demonstrate that q(n,S) is densed in l(n,S) it is sufficient to prove the following: For every $\varepsilon > 0$ and every vertex u of l(n,S) there exists q $\in$ q(n,S) such that $\|q-u\| < \varepsilon$.

Where $\|q-u\|$ is the distance, in R(n,S), between q and u. For suppose that p $\in$ l(n,S) is arbitrary then we have a representation p = $\Sigma\lambda_\nu u^\nu$ where the sum is taken over all vertices u^ν of l(n,S), $\lambda_\nu \geq 0$ and $\Sigma\lambda_\nu = 1$. Let $\varepsilon > 0$ we have to show that there is q $\in$ q(n,S) such that $\|q-p\| < \varepsilon$. But we know that for every index ν there is $q^\nu \in$ l(n,S) such that $\|q^\nu-u^\nu\| < \varepsilon$. Put q = $\Sigma\lambda_\nu q^\nu$ then indeed q $\in$ q(n,S), since q(n,S) is convex, and $\|q-p\| \leq \Sigma\lambda_\nu \|q^\nu-u^\nu\| < \varepsilon$.

For reasons of convenience I shall prove the claim for the supremum norm in R(n,S) that is, the norm:

$$\|x\| = \max \{|x_i|, |x_{ij}|; 1 \leq i \leq n , \{i,j\} \varepsilon S\}$$

This makes no difference since R(n,S) is finite dimensional and thus the supermum norm and the Euclidean norm are equivalent.

We shall proceed by induction on $n \geq 2$. For n = 2 we have two possibilites either S = ϕ or S = {{1,2}}. For S = ϕ the case is trival. For S = {{1,2}}, l(n,S) has   five vertices (0,0,0), (1,0,0), (0,1,0), (1,1,1), (1,1,0).   The first four vertices are classical truth values, they are u^ε for $\varepsilon = (0,0), (1,0), (0,1), (1,1) \in \{0,1\}^2$ respectively. Since c(n,S) $\subseteq$ q(n,S) we have

$u^r \in q(2,S)$ in that case, and no problem arises here. As for the fifth vertex

$u = (1,1,0)$. Let H be a two dimensional space and ψ_1, ψ_2 two orthogonal unit vectors in H. Let $W = E_{\psi_1}$, and for $0 < \theta \leq \pi/2$. Let E_1 be the projection on the space spanned by $(\cos \theta)\psi_1 + (\sin \theta)\psi_2$ and E_2 the projection on the space spanned by $(\cos \theta)\psi_1 - (\sin \theta)\psi_2$ then $E_1 \wedge E_2 = 0$ and we have

$p_1 = \text{tr}[WE_1] = \cos^2\theta$, $p_2 = \text{tr}[(WE_2)] = \cos^2\theta$ and $p_{12} = \text{tr}[W(E_1 \wedge E_2)] = 0$.

Hence $p = (\cos^2\theta, \cos^2\theta, 0) \in q(n,S)$ for all $0 < \theta \leq \pi/2$. Given $\varepsilon > 0$ we can choose θ such that $\|p-u\| = \sin^2\theta < \varepsilon$. (Note that for $\theta = 0$ we have $E_1 = E_2 = W$ and hence $E_1 \wedge E_2 = W$, and thus we cannot prove by the above method that $u \in q(2,S)$, indeed $u \notin q(2,S)$ as we shall see shortly).

Suppose we have proved the claim for $n-1$, $(n > 2)$, that is, that $q(n-1,S')$ is densed in $l(n-1,S')$ for every set of pairs $S' \subseteq \{\{i,j\}\} \mid 1 \leq i \leq n-1\}$. Let u be a vertex of $l(n,S)$ for some fixed set of pairs $S \subseteq \{\{i,j\} \mid 1 \leq i < j \leq n\}$. For a fixed $1 \leq i \leq n$, let u^i be obtained from u by the following definition

$u_i^i = 0$ $u_j^i = u_j$ for $j \neq i$ $1 \leq j \leq n$, $u_{ik}^i = 0$ whenever $\{i,k\} \in S$, $u_{jk}^i = u_{jk}$ for $j,k \neq i$, $\{j,k\} \in S$. Then $0 \leq u_{jk}^i \leq \min \{u_j^i, u_j^i\} \leq 1$ and thus u^i is also a vertex of $l(n,S)$. Moreover from the induction hypothesis it follows that for every $\varepsilon > 0$ there exists $q^i \in q(n,S)$ such that $\|u^i - q^i\| < \varepsilon$. To see that take u^i and eliminate the coordinates u_i^i and u_{ik}^i for $\{i,k\} \in S$ the resulting vector $\bar{u}^i$ is an element of $q(n-1,S_i)$ for some set of pairs among $\{1,2, ..., n-1\}$. From the induction hypothsis there is $\bar{q}^i \in q(n-1,S_i)$ with $\|\bar{q}^i - \bar{u}^i\| < \varepsilon$. Now let q^i be obtained from

$\bar{q}^i$ by adding zeroes as the coordinates q_{ij}^i and q_{ik}^i. Surely $q^i \in q(n,S)$ (since we can always choose $E_i = 0$) and $\|q^i - u^i\| = \|\bar{q}^i - \bar{u}^i\| < \varepsilon$.

Thus for every $1 \leq i \leq n$, there is $q^i \in q(n,S)$ with $\|q^i - u^i\| < \varepsilon$. Since $q^i \in q(n,S)$ there is a Hilbert space H_i projections $E_1^i, E_2^i, ..., E_n^i \in L(H_i)$ and a state W_i on H_i such that $q_j^i = tr(W_i E_j^i)$ $q_{jk}^i = tr[W_i(E_j^i \wedge E_k^i)]$ for $1 \leq j \leq n$, and $\{j,k\} \in S$ (where in particular $E_i^i = 0$). Now let $H = H_1 \otimes H_2 \otimes ... \otimes H_n$ be the tensor product of the spaces, let $W = W_1 \otimes W_2 \otimes ... \otimes W_n$, then clearly W is semidefinite Hermitian and $tr(W) = 1$. Let

$E_i = E_1^i \otimes E_2^i \otimes ... \otimes E_{i-1}^i \otimes I_i \otimes E_{i+1}^i \otimes ... \otimes E_n^i$, where I_i is the identity on H_i. Put $q_i = tr(WE_i)$ and $q_{ij} = tr[W(E_i \wedge E_j)]$ for $1 \leq i \leq n$ and $\{i,j\} \in S$. We have:

$$q_i = \prod_{k \neq i} q_i^k \qquad 1 \leq i \leq n$$

and from Lemma (3–6)b we have for $\{i,j\} \in S$, $i < j$:

$$q_{ij} = \prod_{k \neq i,j} q_{ij}^k \qquad \{i,j\} \in S$$

The proof is concluded when we notice the following: If $0 \leq a_1 \ ... \ a_m$, $b_1 \ ... \ b_m \leq 1$ and $|a_i - b_i| < \varepsilon$, $1 \leq i \leq m$, then $|\prod_{i=1}^m a_i - \prod_{i=1}^m b_i| < m\varepsilon$. Hence since $u_i = u_i^k \in \{0,1\}$ for all $k \neq i$:

$$|q_i - u_i| = |\prod_{k \neq i} q_i^k - \prod_{k \neq i} u_i^k| < (n-1)\varepsilon \qquad 1 \leq i \leq n$$

and since $u_{ij} = u_{ij}^k \in \{0,1\}$ for all $k \neq i,j$ we have

$$|q_{ij} - u_{ij}| = |\prod_{k \neq ij} q_{ij}^k - \prod_{k \neq ij} u_{ij}^k| < (n-2)\varepsilon \qquad \{i,j\} \in S$$

and thus $\|q-u\| < (n-1)\varepsilon$. This can be accomplished for every $\varepsilon > 0$, and therefore $q(n,S)$ is densed in $l(n,S)$.

In fact we have proved more:

Corrolary (3-7): Theorem (3-4)b,c remains valid even if definition (3-3) is restricted to pure states only, that is even if we consider just those $p \in R(n,S)$ for which $p_i = <\phi,E_i\phi> \qquad 1 \leq i \leq n$, and $p_{ij} = <\phi, (E_i \wedge E_j)\phi>$, $\{i,j\} \in S$.

The proof is essentially the same. The only difference is a slight modification of the proof of theorem (3-4)b which is left to the reader.

We have proved that $q(n,S)$ contains the interior of $l(n,S)$. What about the boundary? Since $c(n,S) \subseteq q(n,S)$ all the vertices u^c of $c(n,S)$, which are also vertices of $l(n,S)$ are elements of $q(n,S)$ that is, $q(n,S)$ contains the classical truth functions. But if u is a vertex of $l(n,S)$ which is not an element of $c(n,S)$ then $u \notin q(n,S)$. This is a consequence of the following simple observation:

If tr(WE$_i$) = 1 and tr(WE$_j$) = 1 then necessarily tr[W(E$_i$ ∧ E$_j$)] = 1. If

u ∈ l(n,S)\c(n,S) is a vertex of l(n,S) then necessarily there are indices i,j such that

u$_i$ = 1, u$_j$ = 1 and u$_{ij}$ = 0 and therefore u ∉ q(n,S). Of course, we have proved

that we can get as clsoe as we want to u by elements of q(n,S), but not quite to

u itself. This will turn out to be a crucial point in our discussion on quantum

logic. To sum up we proved:

Theorem (3–8): If u ∈ l(n,S) is a vertex, u ∉ c(n,S) then u ∉ q(n,S)

3.6 "Superficial" Violations of Classical Probability

Formally the simplest case of a non–classical correlation occures already for n

= 2 and S = {{1,2}}. Consider a source of photons all polarized in the z =

(0,0,1) direction in space. Let ψ be the quantum mechanical wave function

associated with these photons so that W = E$_ψ$ is their state. Let ψ' be the wave

function associated with photons polarized in the x = (1,0,0) direction. Let E$_1$ be

the projection on the one dimensional subspace spanned by the unit vector

(cos θ)ψ + (sin θ)ψ', and E$_2$ the projection on the one dimensional subspace

spanned by the unit vector (cos θ)ψ – (sin θ)ψ'. Then E$_1$ ∧ E$_2$ = 0,

p$_1$ = tr(WE$_1$) = cos^{2}θ, p$_2$ = tr(WE$_2$) = cos^{2}θ, p$_{12}$ = tr[W(E$_1$ ∧ E$_2$)] = 0.

Operationally E_1 corresponds to the property "poton is polarized in the direction $a_1 = (\sin\theta, 0, \cos\theta)$" and this corresponds to an experiment in which a polarizer is located in front of the source, oriented in the a_1 direction, $p_1 = \cos^2\theta$ is the frequency of photons which pass through the polarizer. Similarily E_2 corresponds to the property "photon is polarized in the direction $a_2 = (-\sin\theta, 0, \cos\theta)$". $E_1 \wedge E_2 = 0$, hence $E_1 \wedge E_2$ corresponds to the "absurd property" of "null experiment."

Now for a sufficiently small θ, $p_1 + p_2 - p_{12} = 2\cos^2\theta > 1$ so that $p = (p_1, p_2, p_{12}) = (\cos^2\theta, \cos^2\theta, 0)$ is not an element of $c(2,S)$.

This state of affairs already causes trouble for the realist, albeit not an insurmountable trouble. Suppose that we think of the source as an "urn" containing many photons, some of which have property E_1, some have property E_2, and none has both. Suppose furthermoe that we want to maintain that the frequencies p_1, p_2, p_{12} reflect the distribution of these properties in the source. Since $p \notin c(2,S)$ this could not be the case, and therefore something must be wrong with our naive realistic assumption.

There are various ways out of this specific dilemma which save the realistic picture. With one such approach, "hidden variables" we shall deal briefly at the end of this chapter (Section 3.10) and in great detail in chapter 5. For the time being consider another (actually, not quite another) response: The relation $E_1 \wedge E_2 = 0$ can be interpreted in two ways (a) No particle in the source has property E_1 and property E_2. (b) No experiment exists which detects the simultaneous existence of property E_1 and property E_2. Both interpretations seem to be consistent with quantum theory. The first interpretation leads us to trouble, as we have seen. The second interpretation provides a way out. Suppose that "in reality" the source contains many photons which have property E_1 and property E_2, nevertheless we have no way to detect them. In this case $p_{12} = 0$ relfects our inability to perform certain measurements (epistemic problem) but not the actual state of affairs, since "in reality $p_{12} \neq 0$." This solution is an illusion, but it takes more complex cases to demonstrate that.

3.7   Violation of the Clauser-Horne Inequalities

My aim here is to recover Clauser and Horne's (1974) argument in terms of the vocabulary developed previously. In the next section I shall discuss Bell's (1964) original argument, which, from my point of view, is essentially the same; though slightly more cumbersome. I shall consider cases of electron polarization by Stern–Gerlach magnets and ignore experimental difficulties completely. The argument could equally well be applied to photons. Some references to actual experiments appear in the "Notes and Remarks" section. Let H be the two dimensional complex Hilbert space of an electron spin–states. J_x will denote the operator corresponding to a measurement of the spin component of the electron in the x direction, ψ_{+x}, ψ_{-x} are the (normalized) eigenstates of J_x corresponding to "spin up" and "spin down" in the x direction respectively, E_{+x}, E_{-x} the projections on the one dimensional subspaces spanned by ψ_{+x}, ψ_{-x} respectively. We have $J_x = (1/2)(E_{+x} - E_{-x})$. (Throughout we take $\hbar=1$.) Given two electrons the space of their spin states is $H \otimes H$ and the eigenstate of total spin zero is the singlet vector ψ_S. Since the singlet vector is rotationally invarient we can represent it as

$$\psi_S = \frac{1}{\sqrt{2}}[\psi_{+\xi} \otimes \psi_{-\xi} - \psi_{-\xi} \otimes \psi_{+\xi}] \qquad (3\text{--}11)$$

for an arbitrary direction ξ. Let $W_S = E_{\psi_s}$ be the projection in $H \otimes H$ on the one dimensional space spanned by ψ_S that is, W_S is the singlet state. As usual we shall consider a source of electron pairs in the singlet state. When a pair is emitted the two electrons travel in opposite directions "left" and "right". Let x, y, z, w be four arbitrary directions in space and consider the following projections in $H \otimes H$:

$$
\begin{aligned}
E_1 &= (E_{+x} \otimes E_{+x}) \vee (E_{+x} \otimes E_{-x}) \\
E_2 &= (E_{+y} \otimes E_{+y}) \vee (E_{+y} \otimes E_{-y}) \\
E_3 &= (E_{-z} \otimes E_{+z}) \vee (E_{+z} \otimes E_{+z}) \\
E_4 &= (E_{-w} \otimes E_{+w}) \vee (E_{+w} \otimes E_{+w})
\end{aligned}
\qquad (3\text{--}12)
$$

It is easy to identify the properties associated with these projections. Let $\psi \in H \otimes H$ then $E_1\psi = \psi$ if and only if ψ is an eigenvector of the operator $J_x \otimes I$ corresponding to the eigenvalue $+1/2$ (where I is the identity on H). Hence a pair of electrons has the property E_1 if and only if the left electron in the pair has spin "up" in the x direction. The existence of this property is measured by putting a Stern–Gelrach magnet, oriented in the x direction, on the left side of the source. In a similar way we can identify all properties:

property E_1 – The left electron has spin up in the x direction.
property E_2 – The left electron has spin up in the y direction.
property E_3 – The right electron has spin up in the z direction.
property E_4 – The right electron has spin up in the w direction.

Let n = 4 and S = {{1,3}, {1,4}, {2,3}, {2,4}} we want to identify the projections $E_i \wedge E_j$ for {i,j} ∈ S. Since {ψ_{+x}, ψ_{-x}} is a basis for H and so is {ψ_{+z}, ψ_{-z}} it is easy to see that $E_1 \wedge E_3$ is the projection on the one dimensional space spanned by $\psi_{+x} \otimes \psi_{+z}$ that is $E_1 \wedge E_2 = E_{+x} \otimes E_{+z}$. Hence the projection $E_1 \wedge E_3$ corresponds to the property "the left electron has spin up in the x direction and the right electron has spin up in the z direction." The existence of this property is measured by putting a Stern Gerlach magnet, oriented in the x direction to the left of the source and another magnet, oriented in the z direction, on the right hand side of the source. To sum up:

$$E_1 \wedge E_3 = E_{+x} \otimes E_{+z}$$
$$E_1 \wedge E_4 = E_{+x} \otimes E_{+w}$$
$$E_2 \wedge E_3 = E_{+y} \otimes E_{+z} \qquad (3-13)$$
$$E_2 \wedge E_4 = E_{+y} \otimes E_{+w}$$

To obtain the value of $p_i = tr[W_s E_i]$, remember that W_s is the singlet state, which is rotationally invariant hence from (3–11) we get $p_i = 1/2$, $1 \leq i \leq 4$. Consider $p_{13} = tr[W_S(E_1 \wedge E_3)]$. Suppose that (r, θ, ϕ) is a spherical coordinate system in physical space such that z = (1,0,0), then if x = (1, θ, ϕ) we have:

$$\psi_{+x} = e^{(1/2)i\varphi} \cos(\tfrac{\theta}{2})\psi_{+z} + e^{-(1/2)i\varphi} \sin(\tfrac{\theta}{2})\psi_{-z}$$

therefore $p_{13} = \text{tr}[W_3(E_{+x} \otimes E_{+z})] = 1/2 \sin^2(\tfrac{\theta}{2})$ hence we have

$$p_{13} = \frac{1}{2} \sin^2(\frac{\widehat{xz}}{2})$$

$$p_{14} = \frac{1}{2} \sin^2(\frac{\widehat{xw}}{2})$$

(3-14)

$$p_{23} = \frac{1}{2} \sin^2(\frac{\widehat{yz}}{2})$$

$$p_{24} = \frac{1}{2} \sin^2(\frac{\widehat{yw}}{2})$$

where $\widehat{xz}$ is the angle between x and z, and so on.

For n = 4, S = {{1,3}, {1,4}, {1,3}, {2,4}}, the polytope c(n,S) is the Clauser Horne polytope (Section 2.5). We can easily see that for certain choices of x,y,z,w we have p = $(p_1, p_2, p_3, p_4, p_{13}, p_{14}, p_{23}, p_{24}) \notin c(n,S)$. For example, let x,y,w be coplanner directions with $\widehat{xy} = \widehat{yw} = \widehat{xw} = 120°$ and let z = y, then:

$$p = (\frac{1}{2}, \frac{1}{2}, \frac{1}{2}, \frac{1}{2}, \frac{3}{8}, \frac{3}{8}, 0, \frac{3}{8})$$

and

$$p_{13} + p_{14} + p_{24} - p_{23} - p_1 - p_4 = \frac{3}{8} + \frac{3}{8} + \frac{3}{8} - 0 - \frac{1}{2} - \frac{1}{2} = \frac{1}{8} > 0$$

so that inequality (2-15) is violated, and p $\notin$ c(n,S).

Note that so far we have made no use of the principle of locality. Indeed the violations of Bell-type inequalities, in themselves, have nothing to do with the principle of locality. It is only when we try to explain quantum statistics by making extra "realistic" assumptions that the principle of locality enters the scene.

The above result, as the simpler "superficial" case of the prevsious section, poses a problem for the realist. We have domonstrated that p ∉ c(n,S) and therefore we cannot explain the statistical outcome by assuming that the source is an "urn," containing electron pairs in the singlet state, such that the distribution of the properties E_1, E_2, E_3, E_4 in this "urn" is fixed before the measurement. Here there is no easy way out of the dilemma. We cannot redefine the "operational" meaning of the projections E_1, E_2, E_3, E_4. These projections and the joints $E_1 \wedge E_3$, $E_1 \wedge E_4$, $E_2 \wedge E_3$, $E_2 \wedge E_4$ correspond to well defined measurements. Of course one could still provide more sophisticated realistic interpretations (or excuses). We shall consider a few such attempts along the way.

3.8 Violation of Bell Inequalities

With the notation of the previous section consider a source of electron pairs in the singlet state and for three directions x,y,z the projections:

$$
\begin{aligned}
E_1 &= (E_x \otimes E_x) \vee (E_x \otimes E_{-x}) \\
E_2 &= (E_y \otimes E_y) \vee (E_y \otimes E_{-y}) \vee (E_{-y} \otimes E_{-y}) = (E_{-y} \otimes E_y)^{\perp} \\
E_3 &= (E_{-z} \otimes E_{-z}) \vee (E_z \otimes E_{-z})
\end{aligned}
\qquad (3\text{-}15)
$$

As before E_1 is the property "the left electron has spin up in the x direction" and E_3 is the property "the right electron has spin down in the z direction." As for E_2 note that $E_y \otimes E_y$, $E_y \otimes E_{-y}$, $E_{-y} \otimes E_y$ pairwise commute, in fact the comentator of each pair is zero, so that:

$$E_2 = E_y \otimes E_y + E_y \otimes E_{-y} + E_{-y} \otimes E_{-y}$$

and the eigenstates of E_3 have the form

$$\psi = a(\psi_y \otimes \psi_y) + b(\psi_y \otimes \psi_{-y}) + c(\psi_{-y} \otimes \psi_y)$$

where a, b, c are complex constants.

Therefore, in order to check whether an electron pair has the property E_2 we have to locate Stern Gerlach magnets oriented in the y direction on both sides of the source. The pair will have property E_2 if and only if the left electron has spin up in the y direction or the right electron has spin down in the y direction. (Note that this conclusion is independent of the identification of the symbol "∨" as a logical "or" in general. We do not have to make this general assumption since, in this particular case, the projections $E_y \otimes E_y$, $E_y \otimes E_{-y}$, $E_{-y} \otimes E_{-y}$ pairwise commute.)

Now $E_1 \wedge E_2 = E_x \otimes E_{-y}$ $E_1 \wedge E_3 = E_x \otimes E_{-z}$ $E_2 \wedge E_3 = E_y \otimes E_{-z}$,

so that $p_i = \mathrm{tr}(W_s E_i) = 1/2$, $p_{12} = \mathrm{tr}[W_s(E_1 \wedge E_2)] = (1/2)\,\cos^2(\frac{\widehat{xy}}{2})$

$p_{13} = \mathrm{tr}[W_s(E_1 \wedge E_3)] = (1/2)\,\cos^2(\frac{\widehat{xz}}{2})$ $p_{23} = \mathrm{tr}[W_s(E_2 \wedge E_3)] = (1/2)\,\cos^2(\frac{\widehat{yz}}{2})$.

Let $n = 3$, $S = \{\{1,2\}, \{1,3\}, \{2,3\}\}$ then $c(3,S)$ is the Bell–Wigner polytope. It is easy to see that for some choices of x,y,z we have

$p = (p_1, p_2, p_3, p_{12}, p_{13}, p_{23}) \notin c(3,S)$ for example if x,y,z are coplanner with

$\hat{xy} = \hat{xz} = \hat{yz} = 120°$ then $p = (\frac{1}{2}, \frac{1}{2}, \frac{1}{2}, \frac{1}{8}, \frac{1}{8}, \frac{1}{8})$ and thus:

$$p_1 + p_2 + p_3 - p_{12} - p_{13} - p_{23} = 1\frac{1}{8} > 1$$

and inequality (2-12) is violated.

3.9 More General Violations of Classical Constraints

Let S be the set of pairs among 1,2,3, ... n, n+1, ... 2n given by

$S = \{\{i, n+j\} \mid 1 \le i \le n, 1 \le j \le n\}$. Let $x_1, ..., x_n$ be n distinct directions in space and consider the projections in the two electron spin space given by:

$$E_i = (E_{+x_i} \otimes E_{+x_i}) \vee (E_{+x_i} \otimes E_{-x_i}) \quad 1 \le i \le n$$

$$E_{j+n} = (E_{-x_j} \otimes E_{+x_j}) \vee (E_{+x_j} \otimes E_{+x_j}) \quad 1 \le j \le n$$

As before let W_s be the singlet state, then $p_i = tr(W_s E_i) = 1/2$, $1 \le i \le n$,

$p_{n+j} = tr(W_s E_{n+j}) = 1/2$ and for $\{i, n+j\} \in S$,

$$p_{i,n+j} = tr[W(E_i \wedge E_{n+j})] = (1/2)\sin^2(-\frac{\theta_{ij}}{2}) \text{ where } \theta_{ij} \text{ is the angle between } x_i \text{ and}$$

x_j.

Suppose that $p = (p_1, p_2, \cdots p_n, p_{n+1}, \cdots p_{2n}, \cdots p_{ij} \cdots) \in c(2n,S)$ then by theorem (2-3) there is a probability space (X, Σ, μ) and events $A_1, A_2 \cdots A_n, A_{n+1}, \cdots, A_{2n} \in \Sigma$ such that $p_i = \mu(A_i)$ $i = 1, 2 \cdots, 2n$ and $p_{i,n+j} = \mu(A_i \cap A_{n+j})$ for $\{i, n+j\} \in S$. In particular for all $1 \leq i \leq n$ we have $A_i = \tilde{A}_{n+i}$ and $p_{i,n+i} = \mu(A_i \cap A_{n+i}) = 0$. Hence we have

$$\mu(A_i \cap A_j) = \mu(A_i) - \mu(A_i \cap \tilde{A}_j) = \mu(A_i) - \mu(A_i \cap A_{n+j}) = p_i - p_{i,n+j} =$$

$(1/2)\cos^2(\theta_{ij}/2)$. But inequality (2-19) must be satisfied, so we conclude that $p \in c(2n,S)$ only if

$$\sum_{i=1}^{n} p_i - \sum_{1 \leq i < j \leq n} (p_i - p_{i,n+j}) = \sum_{i=1}^{n} \mu(A_i) - \sum_{1 \leq i < j \leq n} \mu(A_i \cap A_j) \leq 1 .$$

Substituting the values obtained above we get: $p \in c(2n,S)$ only if

$$(n/2) - (1/2) \sum_{1 \leq i < j \leq n} \cos^2(\theta_{ij}/2) \leq 1 \tag{3-16}$$

We use the identity $\cos^2(\theta_{ij}/2) = (1/2)\cos\theta_{ij} + (1/2)$ and remember that $\cos\theta_{ij} = x_i \cdot x_j$ that is, $\cos\theta_{ij}$ is just the scalar product of the unit vectors x_i and x_j. Substituting in (3-16) and rearranging the components we get that $p \in c(2n,S)$ only if

$$\sum_{1 \leq i < j \leq n} x_i \cdot x_j \geq -\frac{1}{2}(n^2 - 5n + 8) \qquad\qquad (3\text{--}17)$$

Let $v = \sum_{i=1}^{n} x_i$ then $\sum_{1 \leq i < j \leq n} x_i \cdot x_j = \frac{1}{2}(v \cdot v - n)$. Therefore $p \in c(2n,S)$ only if:

$$v \cdot v = \| \sum_{i=1}^{n} x_i \|^2 \geq -n^2 + 6n - 8 \qquad\qquad (3\text{--}18)$$

$\|\sum x_i\|^2$ is non negative; the real function $f(t) = -t^2 + 6t - 8$ is positive in the interval $2 < t < 4$, and obtains its maximum at $t = 3$. Therefore a violation of (3–18) may occur for $n = 3$. In this case we must have by (3–18):

$\|x_1 + x_2 + x_3\|^2 \geq 1$. But if we choose x_1, x_2, x_3 to be coplanner with $\theta_{12} = \theta_{13} = \theta_{23} = 120°$ then $\|x_1 + x_2 + x_3\|^2 = 0$. This is the maximal violation of inequalities (2– 18), (2–19) in an E.P.R. set–up.

3.10   Preliminary   Discussion   of   the   Results,   Bohrs   Views   and Antirealism

When we attempt to explain and justify the axioms of probability theory we very often use the "balls in the urn" model. Indeed this model captures a fundamental intuition regarding probability: The frequency of red balls in a random sample of balls approximates the proportion of red balls in the urn. More generally, observable frequencies reflect a distribution of properties which exists independently of our methods of random sampling and measurement. The "balls in the urn" view is adopted by almost every elementary textbook on probability.

Some approaches to probability take a different starting point. The subjectivist school maintains that the probability of a proposition is a measure of our degree of belief in that proposition. According to this view the axioms of classical probability are justified because they reflect the rules of propositional logic. (For example, a rational agent will not assign probability 3/4 to two conflicting propositions).

Both the objectivist "balls in an urn" metaphor and the subjectivist view lead to the same set of constraints on correlations, namely the facet inequalities of $c(n,S)$. The violations of these constraints by quantum frequencies thus poses a major problem to *all* schools of classical probability. I take this fact to be the major source of difficulty which underlies the interpretation of quantum theory. For this reason Feynman "can safely say that nobody understands quantum mechanics."

Those who maintain that the axioms of probability theory reflect the rules of propositional logic, naturally come to suspect those rules. This is the source of quantum logic, which will be discussed in great detail in the next chapter.

The more common view of probability – at least among physicists – is the objectivist one. For the objectivist the violation of the classical constraints mean that we can no longer hold the "balls in an urn" metaphor without additional hypotheses. Therefore an alternative to the traditional approach has to be found. This alternative can take as its starting point a metaphysical thesis or some extra physical assumptions. We shall briefly discuss two metaphysical approahces below, and in the next section introduce the physical "hidden variable" approach.

The Copenhagen Interpretation: The failure of the "balls in an urn" model indicates, according to this school, a breakdown of our classical concept of "physcial property," and more generally, the inadaquacy of our classical notion of "physical magnitude." We can ascribe a momentum value to a particle only in conjunction with a specific experiment. Similarily, we can ascribe a spin value in a given direction to a particle, only in the context of a specific experiment. Micorscopic properties, properly speaking, are not properties of the individual microscopic system involved. They are, rather, a reflection of the entire experimental set-up. It follows that the frequency of particles (or systems of particles), having a certain property, *does not* mirror a distribution which had existed prior to the experiment. Speaking about microscopic properties independently of experiment, makes no sense at all. No wonder that classical constraints are violated.

Let as examine this view more carefully in the context of the violation of Clauser-Horne inequalities described in Section (3-7). According to the classical view the source of electron pairs in the singlet state is like an "urn" containing a vast number of electron pairs. Some such pairs have the property corresponding to the projection E_1, namely the left electron in the pair has spin up in the x direction. Some pairs have the property E_2 and so forth. The values of

$p = (p_1, p_2, p_3, p_4, p_{13}, p_{14}, p_{23}, p_{24})$ should, according to the classical view, reflect this prior distribution of pairs. Since p is not classical we face a dilemma.

But as far as Bohr is concerned E_1 is not a property of a pair of electrons which can be defined independently of an experiment. We can say that an electron pair "has property E_1" only if we actually perform a measurement by putting a Stern Gerlach magnet oriented in the x direction to the left of the source and observe that the left electron goes "up". Similarily we can maintain that the electron pair "has property $E_1 \wedge E_3$" only when we put Stern–Gelrach magnets, one oriented in the x direction to the left of the source and one oriented in the z direction to the right of the source, and observe that both electrons go "up".

Thus the properties E_1, $E_1 \wedge E_3$, and so forth, are necessarily associated with particular experimental arrangements. Since, however, the experimental arrangements for each one of the properties E_1, E_2, ..., $E_1 \wedge E_3$, $E_1 \wedge E_4$, ... are different, there is no a–priori reason to think that the observed frequency vector p will satisfy classical constraints, and indeed sometimes it does not.

I think that Bohr's approach is essentially realistic. Bohr does not deny the existence of particles (though particles are not the classical "little balls") nor does he deny the existence of their properties. It is only the ascription of "intrinsic" properties to particles, properties which are independent of the particular context, which is critisized by Bohr. In fact we face a somewhat similar situation in the macroscopic world. We say that the table is brown, but the color of an object is not an intrinsic property, it varies with the external illumination conditions. We can avoid this difficulty by speaking of the color of the table in "normal illumination conditions," and so forth.

The major difference between this classical case and the quantum case is that in the latter we cannot define "normal conditions", in which we can determine *all* the physically relevant properties simultaneously. There is no single experimental set-up in which we can measure the position and the momentum of the particle at once, or measure two of its different spin components simultaneously. Thus, we can either choose a momentum-space description, or a configuration-space description, but cannot have both. The two descriptions are "complementary".

Note that not all quantum properties have this disturbing nature. The electric charge, for example, and in fact all magnetudes for which a superselection rule applies, are truely intrinisic magnetudes in the sense that they are not complementary to any other physical observable.

The Antirealist View: While Bohr's "solution" takes a revisionist attitude towards properties, the antirealist attacks the concept of cause. His, in fact, is an argument in the sceptical tradition which originated with Hume. An argument of this nature has been proposed, e.g. by van Fraassen (1982). I shall not repeat van Fraassens reasoning here but rather adopt a more general perspective.

Hume scepticism regarding causal relations is well known. We observe that an event of type A is always followed by an event of type B. As a result of this regularity we form the opinion that A is the cause of B. But the ascription of causal connection between A and B is in the eye of the beholder and not necessarily a feature of the world. It is a human habit to form such an opinion.

If scepticism with respect to causal connections between observable events is justified, all the more so scepticism with respect to stipulated hidden causes. We maintain that an object falls to the ground as a result of gravity. For the realist the gravitational field does not merely function as an explanation of the phenomenon, it is its actual cause. But we cannot observe gravitational fields, only their effects. A Humian sceptic will surely be suspicious with respect to such stipulated causes.

Thus, according to Hume, we ascribe causal relations among observable events, and to those events, for which no apparent cause exists, we stipulate a hidden cause. Since these casusal connections are human ascriptions they are not necessarily real. If this is the case, how can we safely maintain that the set of all hidden causes is at all consistent? Maybe there is no way to ascribe hidden causes to all phenomena in a logically consistent manner?

Quantum mechanics, maintains the sceptic, demonstrates that no consistent set of causes is possible. If we want to hold to the view that a particle goes "up", in a Stern-Gerlach magnet, *because* it has spin up in the relevant direction (that is, because it had positive spin prior to entering the magnetic field) then we should expect frequencies to satisfy classical constraints. Since they do not, our conclusion is that what we call "spin up in the x direction" is not the cause of the phenonmon, but only its *name*.

Quantum theory predicts sucessfully a vast number of verified results. This impressive success rate is not due to the fact that quantum mechanics is a theory of causes. There are no causes, there are only the phenomena themselves.

In a world without causes "anything goes". The observed frequencies need not satisfy any constraints, apart from those which reflect pure conventions (for example: frequancy is, by its defintion, a number between zero and one). Indeed there is no a–priori reason to believe that frequencies converge to a limit value in the long run; frequencies may very well oscilate and diverge. The violation of the classical constraints, in an Einstein–Podolsky–Rosen type of experiment, comes as no great surprise to the antirealist.

3.11   Hidden Variables and the Principle of Locality

The "hidden variable" approach is the exact opposite of the antirealist view, both in content and in spirit. It is firstly an attempt to provide a physical, rather than a metaphysical solution to the conceptual difficulties underlying quantum statistics and secondly, it is realistic in its ideology. The source of trouble, according to the "hidden variable" school lies in the fact that quantum theory provides only a partial and fragmented account of physical reality.

In the introductory chapter we have already considered the idea that there may exist a theory, which extends quantum mechanics in that it tells us what precisely happends to the momentum of a particle, during a position measurement and vice–versa. Such a theory can be made compatible with the uncertainty principle, since it postulates that a certain well defined disturbance occurs during a position measurement, a disturbance whose precise magnitude can be computed from the value of the initial conditions. Let us formulate this idea in the context of the Hilbert space formalism.

With every physical system we shall associate a set of hidden variables call it Λ. The elements of Λ can be scalars, vectors, functions or any type of mathemtical objects suitable for our purpose. Each element $\lambda \in \Lambda$ will be called "a hidden variable state." Given a fixed hidden variable state $\lambda \in \Lambda$, we can calculate from it the simultaneous value of all physical observables associated with the system, even those whose simultaneous value is undefined, according to quantum theory.

Given a quantum property of the system (for example, the property "spin up in the x direction" for an electron), we can associate with it a set $A \subset \Lambda$ of all the hidden variable states λ, such that if the system is in the state λ, then it has the said property (for example $\lambda \in A$ if and only if an electron in the state λ has spin $+1/2$ in the x direction).

Suppose H is the Hilbert space associated with the system. Then with every projection $E \in L(H)$ we associate a subset $A(E) \subset \Lambda$ such that the system has the property associated with E if and only if its hidden varible state λ is in the set $A(E)$.

Suppose moreover that the quntum state of the system is given by the statistical operator W. Then with W there corresponds a probability measure μ, on a suitable algebra of subsets of Λ, such that $\text{tr}[WE] = \mu[A(E)]$ for all $E \in L(H)$. In this way, our "hidden variable theory" is capable of determining the simultaneous value of all quantum observables, and to recover the details of quantum statistics. In such a theory the quantum mechanical states are reduced to probability measures on Λ. Hence the relation between quantum mechanics and the hidden variable theory is like the relation between statistical thermodynmics and classical mechanics.

Notice however, that we *cannot* have $A(E_1 \wedge E_2) = A(E_1) \cap A(E_2)$ for all E_1, $E_2 \in L(H)$. The reason is obvious, if the above relation holds then, by theorem (2–3), all quantum correlations would have been classical. Hence we sometimes have $A(E_1 \wedge E_2) \neq A(E_1) \cap A(E_2)$. This is nevertheless a reasonable assumption. The experimental determination of the property E_1 may "disturb" the property E_2 and vice versa. $A(E_1 \wedge E_2)$ is the set of all hidden variable states λ such that a system has property $E_1 \wedge E_2$ if and only if $\lambda \in A(E_1 \wedge E_2)$. This means that $\lambda \in A(E_1 \wedge E_2)$ if and only if a direct measurement of $E_1 \wedge E_2$ on a system in the hidden variable state λ, will give positive result. But the measurement of $E_1 \wedge E_2$ involves an experiment which is different from the one involved in the determination of E_1 alone, or the determination of E_2 alone. Thus we may have for example $\lambda \in \Lambda$, such that a direct measurement of E_1 on a system in the hidden variable state λ will give a positive result, and a direct measurement of E_2 will also give a positive result, but a direct measurement of $E_1 \wedge E_2$ will give a negative result. Even this simple and unsophisticated scheme is already problematic. The trouble lies not so much with quantum mechanics, such as scheme can easily be made compatible with quantum theory. Our schematic hidden variable theory, nevertheless, conflicts with the special theory of relativity.

In order to see why consider the Bell–Wigner version of Einstein, Podoslay, Rosen experiment in Section 3.8   There we have considered the case $n = 3$, and $S = \{\{1,2\}, \{1,3\}, \{2,3\}\}$. We have taken electron pairs in the singlet state W_s and three properties E_1, E_2, E_3 given in formula (3–15). We obtained

$$p_i = \text{tr}[W_s E_i] = 1/2 \quad i = 1,2,3$$

$$p_{ij} = \text{tr}[W_s(E_i \wedge E_j)] = 1/8 \quad 1 \leq i < j \leq 3$$

Suppose we adopt the hidden variable theory above then we have sets $A(E_i)$ $i = 1,2,3$ and $A(E_i \wedge E_j)$, $1 \leq i < j \leq 3$, and a measure μ on the set of hidden variable states, such that $\mu(A(E_i)) = 1/2$ $i = 1,2,3$, $\mu(A(E_i \wedge E_j)) = 1/8$ for $1 \leq i < j \leq 3$. But $p = (p_1, p_2, p_3, p_{12}, p_{13}, p_{23}) \notin c(3,S)$ and hence, by theorem (2–3) we must have $A(E_i \wedge E_j) \neq A(E_i) \cap A(E_j)$ for some pair of indices $1 \leq i < j \leq 3$. Suppose, without loss of generality, that $i = 1$, $j = 3$, and moreover that there exists $\lambda_0 \in \Lambda$ such that $\lambda_0 \in A(E_1) \cap A(E_3)$ but $\lambda_0 \notin A(E_1 \wedge E_3)$ (other cases can be handled in a similar manner).

Now $\lambda_0 \in A(E_1)$ means that an electron pair, in the hidden variable state λ_0, has property E_1 namely, if we put a Stern–Gerlach magnet oriented in the x direction to the left of the source, and put no magnet to the right of the source, the left electron will go "up." Also $\lambda_0 \in A(E_3)$, so if we put a Stern–Gerlach magnet oriented in the z direction to the right of the source, and put no magnet to the left of the source, the right electron will go "down." But $\lambda_0 \in A(E_1 \wedge E_2)$ therefore if we put two Stern–Gerlach magnets, the first oriented in the x direction to the left of the source, and the second oriented in the z direction to the right of the source, then either the left electron will go "down" (with some probability q, say), or the right electron will go "up" (with probability q').

All this means that our hidden variable theory is not covariant. Suppose that we make the distance between the left and the right magnets very large, and suppose moreover that we have a source of electron pairs *all in the hidden variable state* λ_0, then we can transmit superluminal messages from right to left, with very high probability.

On the left hand side we put a Stern–Gerlach magnet oriented in the x direction. As long as the person on the right hand side has no message to deliver she puts no magnet on her side. In this case all the electrons on the left will go "up." If the person on the right wants to deliver a message (a warning, say), she puts a Stern–Gerlach magnet oriented in the z direction on her side. As a result the electrons on the left will go down with probability q. Hence the probability that after N electron, at least one will go "down" on the left, is $1 - (1 - q)^N$, that is almost certainly for a sufficiently large N. Since the distance between the right and left hand sides is very large and since the flux of electrons can be made large as well, the message can be delivered superluminally almost with certainty.

The hidden variable approach, just considered, does not necessarily imply that we can perform such tricks in practice. They may not be a way to isolate all those particle pairs which are in the state λ_0 and use them for superluminal signaling. Nevertheless, the above argument indicates that hidden variables electron pair states, such as λ_0, are not covariant, and thus the entire theory is not covariant. Even if we cannot control the hidden variable states and use them for technological purposes, still, the hidden variable theory itself reduces special relativity to the level of a statistical, rather than a principal theory of space and time.

This is a major difficulty with the hidden variable approach. Sometimes the situation is summarized by the saying: "realism and locality cannot coexist," where by "realism" one means the possibility of ascribing properties to particles (or systems of particles) indpendently of measurement. But this slogan is false, there are theories which are both "realist," in the above sense, and local. We shall discuss the hidden variable approach in greater detail in Chapter 5.

3.12 Notes and Remarks

My analysis of quantum probability follows von Neumann's classic (1955). Theorem (3-2), the uncertainty relations in their full generality, is proved in Messiah textbook (1963). The structure of the lattice of closed subspaces of a Hilbert space if due to Birkhoff & von Neumann (1943). The remarkable theorem

(3–3) is due to A. Gleason (1957), an elementary proof of this theorem is provided in Cooke, Kean and Moran (1985). Gleason proved his theorem in three stages. The first is a tedious, but straightforward reduction of the general case to the three dimensional real Hilbert space $R^{(3)}$. The second part consists in showing that probability measures, on subspaces of $R^{(3)}$, are continuous; this is an elementary argument, which makes an ingenous use of three dimensional geometry. In the last part Gleason proved the claim using the theory of group representations (applied to the rotation group). It is this last section which is less elementary.

Theorem (3–4) on the possible range of values of quantum correlations appeard in a somewhat less general form in Pitowsky (1986).

The Einstein–Podosky–Rosen experiment was proposed by these authors in their 1935 paper, which is one of this century's scientific classics. The experiment proposed by EPR involved position and moemtum values rather than spin or polarization values, and thus it remained a thought experiment. The contemporary version which involves spin was proposed by Bohm (1951). The statistical analysis of this experiment, given in Section 3.8, is essentially due to Bell (1964). Bell's innovative argument was subsequently simplified by Wigner (1970). The verison in 3.7 is due to Clauser and Horne (1974). All these authors do not use projections in Hilbert space explicitely, however. More general violations of the inequality are indicated in the work of Mermin and his students, the references are mentioned in Section 2.9.

The most recent verification of these predictions is in Aspect, Grangier and Roger (1981). For a survey of previous experiments see Clauser and Shimony

(1978). It should be noted that the best results are obtained by examining photon polarization states. It is doubtful whether free electrons could be used for similar purposes, see Mott and Massey (1965).

Bohr (1949) is a good summary of his view and his debate with Einstein. Hume's sceptical arguments concerning causal relations is in the classic masterpiece Hume (1739). The most profound antyrealist account of the quantum mechanical puzzles is due to van Fraassen (1982), see also his book (1980). An analysis which is more in line with the physical, rather than philosophical tradition, is given in d'Espangant (1976). Many hidden variable theories have been proposed along the years. A summary of the most important ones is in Belifante (1973). See also Bub (1977). The version which appears here is the most general one, and is closely connected to the so-called "contextual" hidden variable theories, see Gudder (1970) and Shimony (1984). A more detailed account of hidden variables is given in Chapter 5.

4. Quantum Logic

"Don't seek to have a revolution and minimize it too."

Hilary Putnam

4.1 Quantum Correlation Polytopes and "Truth"

In our discussion of classical correlations we have arrived at the notion of "correlation polytopes" in two different routes. Taking the objectivist approach, we derived the constraints on frequencies, in the form of linear inequalities which are just the facets of c(n,S). Taking the subjectivist line we considered the vertices of c(n,S) that is, all possible truth value assignments to n propositions a_1, a_2, ..., a_n and pairs $a_i \wedge a_j$, $\{i,j\} \in S$. According to the subjectivist view, a correlation vector $p \in c(n,S)$ is just a bet (weighted average) on all those truth functions.

In the quantum mechanical case we have confined ourselves, so far, to the objectivist view. In the previous chapter we derived the constraints on the frequency of quantum events; these are summarized in theorem (3–4) and theorem (3–8). What we lack is the complementary view of quantum correlations that is, the interpretation of the elements $p \in q(n,S)$ in terms of bets on all possible truth value assignment to a set of propositions a_1, ... a_n and pairs $a_i \wedge a_j$, $\{i,j\} \in S$. This is of course no accident since the vertices of q(n,S) (which by theorem (3–8) are not always elements of q(n,S)), are not truth values. If u is such a vertex we may have $u_i = u_j = 1$ and $u_{ij} = 0$ for some $\{i,j\} \in S$.

Let us take a bold step. By analogy with the classical case we shall declare all the vertices of q(n,S) as truth functions. This means that we can have two propositions a_1, a_2 such that a_1 is true, a_2 is true, but $a_1 \wedge a_2$ false!

What is the meaning of this? I shall ignore this question for the time being and proceed in a purely formal fashion. The interpretation (or rather, several possible interpretations) of quantum logic will be discussed in sections 4.4 through 4.7.

Even from a purely formal perspective the new "truth functions" are not entirely analogous to the classical ones. It is easy to see that the vertices of c(n,S) can be understood both as truth functions and also as extreme value of possible frequencies. Thus if n = 2, S = {{1,2}} then the vertex $(p_1, p_2, p_{12}) = (1,0,0)$ of c(2,S) represents a case where all objects under consideration have the first property, none has the second property and thus none has both. In the quantum case the non-classical vertices of q(n,S) are not themselves elements of q(n,S) (theorem (3–8)) thus the vertex $(p_1, p_2, p_{12}) = (1,1,0)$ of q(2,S), for S = {{1,2}}, does not represent possible quantum frequencies. It is impossible to find sources of particles (or systems of particles) such that all have property E_1 all have property E_2 and none has $E_1 \wedge E_2$, though we can *approximate* that case by quantum frequencies to any desireable degree (e.g., (0.99, 0.99, 0) $\in$ q(2,S)). *All this means that our bold decision is an extention of quantum mechanics.* By admitting the vertices of q(n,S) to our story we make an additional assumption. We shall see that, given a realistic interpretation of quantum logic, this additional assumption has *physical* consequences similar to the ones we have found in the analysis of hidden variable theories.

In the following section I shall proceed to develop quantum logic on a purely formal level, in 4.3 I shall introduce the important theorem of Kochen and Specker; following this I shall discuss some possible interpretations.

4.2 Formal Development of Quantum Logic

In classical logic the truth value of a conjunction is the product of the truth values of the conjuncts. In quantum logic the truth value of a conjunction is less or equal to the product of the truth value of the conjuncts. Hence by admitting the new truth functions we lose an important property of classical propositional logic, namely that the truth values of a complex proposition is uniquely determined by the truth value of its atomic constituents. Hence, to define quantum truth values of complex propositions we shall use induction on the length of formulas.

The set of all well formed propositions is defined as in the classical case. If $A = \{a_1, a_2, ..., a_n ...\}$ is the set of atomic propositions then the set of all well formed propositions is defined by induction: (a) Every atomic proposition is a well formed proposition; (b) If b,c are propositions so are ($\sim$b) and (b $\wedge$ c). The other logical connectives are defined as shorthand. Thus (b $\vee$ c) is a shorthand for

~[(~b) ∧ (~c)] and (b → c) is a shorthand for [(~b) ∨ c]. Let W denote the set of all well formed propositions.

<u>Definition (4–1):</u>   A quantum truth function is a function θ: W → {0,1} which satisfies:

(i) $\theta(\sim b) = 1 - \theta(b)$ for all propositions b ∈ W

(ii) $\theta(b \wedge c) = \theta(c \wedge b) \leq \theta(b)\theta(c)$ for all b,c ∈ W

(iii) $\theta(b \wedge b) = \theta(b)$ for all b ∈ W

(iv)   If $\theta'(b) \leq \theta'(c)$ for all quantum truth functions θ' then
        $\theta(b \wedge d) \leq \theta(c \wedge d)$ for all d ∈ W .

The inductive character of the definition is evident from (iv). Note that the number of distinct truth functions, defined for a fixed proposition b, is bounded by 2^k, where k is the number of well formed sub-formulas of b. Note also that θ is a classical truth function if and only if equality holds in (ii) for all b,c ∈ W. A proposition b is said to be a quantum tautology (quantum logical falsity), if $\theta(b) = 1$ ($\theta(b) = 0$ respectivly), for all quantum truth functions θ. Since every classical truth function is, in particular, a quantum truth function, every quantum tautology (logical falsity) is also a classical tautology (logical falsity). Hence there are fewer quantum tautologies and quantum logical falsities then classical ones.

Lemma (4-1): (a) $\theta(a \vee b) \geq \theta(a) + \theta(b) - \theta(a)\theta(b)$

(b) $\theta(a \rightarrow b) \geq 1 - \theta(a) + \theta(a)\theta(b)$

Proof: (a) By definition $\theta(a \vee b) = \theta[\sim[(\sim a) \wedge (\sim b)]] = 1 - \theta[(\sim a) \wedge (\sim b)] \geq$

$1 - \theta(\sim a) \theta (\sim b) = 1 - (1 - \theta(a))(1 - \theta(b)) = \theta(a) + \theta(b) - \theta(a)\theta(b)$

(b) $\theta(a \rightarrow b) = \theta[(\sim a) \vee b)] \geq \theta(\sim a) + \theta(b) - \theta(\sim a)\theta(b) =$

$1 - \theta(a) + \theta(b) - (1 - \theta(a))\theta(b) = 1 - \theta(a) + \theta(a)\theta(b)$

Corrolary (4-2): If $\theta(a) \leq \theta(b)$ then $\theta(a \rightarrow b) = 1$

since $\theta(a) \leq \theta(b)$ entails $1 - \theta(a) + \theta(a)\theta(b) = 1$ and by the previous lemma

$\theta(a \rightarrow b) = 1$.

Theorem (4-3): For all $a,b,c \in W$ the following propositions are quantum tautologies:

(1) $(a \vee (\sim a))$,

(2) $(a \rightarrow (a \vee b))$,

(3) $((a \wedge b) \rightarrow a)$,

(4) $((\sim(\sim a)) \leftrightarrow a)$,

(5) $((\sim(a \wedge b)) \leftrightarrow ((\sim a) \vee (\sim b)))$ (De-Morgan rule),

(6) $(((a \wedge b) \vee (a \wedge c)) \rightarrow (a \wedge(b \vee c)))$

Proof: (1) From lemma (4-1) $\theta(a \vee (\sim a)) \geq \theta(a) + \theta(\sim a) - \theta(a)\theta(\sim a) = 1$ for all quantum truth functions. θ.

(2) $\theta(a \vee b) \geq \theta(a) + \theta(b) - \theta(a)\theta(b)$ therefore $\theta(a \vee b) \geq \theta(a)$ and by corrolary (4-2) $\theta(a \rightarrow a \vee b)) = 1$.

(3) Since $\theta(a \wedge b) \leq \theta(a)\theta(b) \leq \theta(a)$ the claim follows again from corrolary (4-2).

(4) $\theta(\sim(\sim a)) = 1 - (1 - \theta(a)) = \theta(a)$ hence $\theta((\sim(\sim a)) \leftrightarrow a) = 1$ by corrolary (4-2).

(5) Put $\sim\sim a$ as a shorthand for $(\sim(\sim a))$. Since $\theta(\sim\sim a) = \theta(a)$ for all quantum truth functions θ we have, by induction rule (iv) in definition (4-1): $\theta(\sim\sim a \wedge b) = \theta(a \wedge b)$. Similarly $\theta(b) = \theta(\sim\sim b)$. Then by the same induction principle $\theta (b \wedge \sim\sim a) = \theta(\sim\sim b \wedge \sim\sim a)$. Since $\theta(b \wedge \sim\sim a) = \theta(\sim\sim a \wedge b)$ by rule (ii) in definition (4-1), we conclude $\theta(a \wedge b) = \theta(\sim\sim a \wedge \sim\sim b)$. Hence $\theta(\sim(a \wedge b)) = \theta[\sim[(\sim\sim a) \wedge (\sim\sim b)]]$. But by definition of the connective $\vee$ we get: $\sim[(\sim(\sim a)) \wedge (\sim(\sim b))] = [(\sim a)) \vee (\sim b)]$. Hence $\theta(\sim(a \wedge b) = \theta((\sim a) \vee (\sim b))$ and thus $\theta[\sim(a \wedge b) \leftrightarrow ((\sim a) \vee (\sim b)] = 1$ by corrolary (4-2).

(6) The proof of this quantum tautology requires some preparations. Let d,e be two propositions such that $\theta(d) \leq \theta(e)$ for all quantum truth functions θ, then $\theta(d \vee f) \leq \theta(e \vee f)$ for all propositions f. To see that remember that by definition $d \vee f = \sim ((\sim d) \wedge (\sim f))$ and $e \vee f = \sim ((\sim e) \wedge (\sim f))$. Also $\theta(d) \leq \theta(e)$ if and only if $\theta(\sim e) \leq \theta(\sim d)$. Hence by rule (iv) in definition (4-1)

$\theta((\sim e) \wedge ((\sim f)) \leq \theta((\sim d) \wedge (\sim f))$ thus $\theta(\sim((\sim d) \wedge (\sim f))) \leq \theta(\sim((\sim e) \wedge (\sim f)))$ which, by definition of $\vee$ means: $\theta(d \vee f) \leq \theta(e \vee f)$.

Now let d,e be as above, $\theta(d) \leq \theta(e)$ for all θ, then $\theta(e \vee d) = \theta(e)$ for all θ. This is evident since we have proved $\theta(d \vee f) \leq \theta(e \vee f)$ for all f, put $f = e$ and remember that $\theta(e \vee e) = \theta(e)$, by rule (iii) in definition (4–1) and the definition of the connective $\vee$. Hence $\theta(d \vee e) \leq \theta(e)$. Also

$\theta(d \vee e) \geq \theta(e) + \theta(d) - \theta(e)\theta(d) \geq \theta(e)$ hence $\theta(d) \leq \theta(e)$ for all θ entails

$\theta(d \vee e) = \theta(e)$ for all θ.

Now we are ready to prove tautology (6). Since $\theta(b) \leq \theta(b \vee c)$ for all θ we have by rule (iv) $\theta(a \wedge b) \leq \theta(a \wedge (b \vee c))$. Put $d = a \wedge b$ and

$e = (a \wedge (b \vee c))$ and $f = a \wedge c$ then $\theta(d) \leq \theta(e)$ for all θ

entails $\theta(d \vee f) \leq \theta(e \vee f)$ for all θ, as we have shown above. Substituting the values of d,e,f we get $\theta((a \wedge b) \vee (a \wedge c)) \leq \theta[(a \wedge (b \vee c)) \vee (a \wedge c)]$.

Now put $d = a \wedge c$ and $e = a \wedge (b \vee c)$ then $\theta(d) \leq \theta(e)$ for all θ and thus

$\theta(d \vee e) = \theta(e)$ for all θ that is, $\theta[(a \wedge (b \vee c)) \vee (a \wedge c)] = \theta(a \wedge (b \vee c))$.

Hence $\theta((a \vee b) \wedge (a \vee c)) \leq \theta(a \wedge (b \vee c))$ for all θ and therefore the claim follows from corollary (4–2).

The last theorem demonstrates that many classical tautologies are in fact quantum tautologies. An important example of a classical tautology which is not a quantum tautology is given by the reverse implication in (6) the distributirity rule:

$$(a \wedge (b \vee c)) \rightarrow ((a \wedge b) \vee (a \wedge c)) \qquad (4-1)$$

To see that the distributivity rule is not a quantum tautology take a,b,c to be atomic propositions and θ to be a quantum truth function such that

$\theta(a) = 1$, $\theta(b) = \theta(c) = 0$, $\theta(a \wedge b) = \theta(a \wedge c) = 0$, $\theta((a \wedge b) \vee (a \wedge c)) = 0$, all these values are classical. But in opposition to the classical case we can choose

$\theta(b \vee c) = 1$ and $\theta(a \wedge (b \vee c)) = 1$ and hence

$0 = \theta((a \wedge b) \vee (a \wedge c)) \neq \theta(a \wedge (b \vee c)) = 1$.

In section (3.3) we have briefly discussed Birkhoff and von Neumann (1942) proposal to identify the closed subspaces of a Hilbert space as "prpositions", or rather, equivalence classes of propositions, and to identify the lattice operations as logical connectives. This proposal, and many others which followed, suffer from two disadvantages:

(1) Generally speaking, equating an algebraic structure with "logic" is problematic. The closed subspaces are supposed to be the equivalence classes of propositions, analogous to the elements of the Lindenbaum–Tarski algebra of classical logic; but it is not clear what is the relevant notion of "logical equivalence" which brings about the Hilbert space lattice structure, as opposed to a Boolean structure. In the "algebraic" approach, logic seems to be divorced from the notion of "truth".

(2) Quantum logic is supposed to be justified by empirical consideration, but the algebraic structure of the Hilbert lattice is not directly connected with physical phenomenology. The relation between Birkhoff and von Neumann's version and the experimental results is only indirect. One requires the entire theoretical machinary of quantum mechanics to justify quantum logic.

The approach adopted here overcomes these difficulties. On the formal level it is directly related to propositions, rather than equivalence classes. It also directly follows from experimental results, and not from their theoretical explanations. Our starting point has been the violation of the classical constraints by quantum-frequencies, a fact which has been varified by numerous experiments.

The rules given in definition (4-1) are not sufficient to generate all the relations which exist among closed subspaces of a Hilbert space. If we define the equivalence class of a proposition a to be:

$$[a] = \{b \in W \mid \theta(b) = \theta(a) \text{ for all quantum truth values } \theta\}$$

and the operations among equivalence classes in the usual manner (e.g. [a] ∨ [b] = [a ∨ b]). The resulting algebraic structure is, of course, not isomorphic to the lattice of closed subspaces of a Hilbert space, it is not even atomic. In that sense definition (4-1) tells only part of the story. It is true nevertheless that the lattice of closed subspaces of a Hilbert space is a model for our version of quantum logic since rules (i) – (iv) are clearly satisfied. I suspect that no finite or even recursive extention of the set of rules (i) – (iv) will suffice to derive *all* the geometric relations which exist among the closed subspaces of a complex or real Hilbert space. (For otherwise we would be able to obtain a purely algebraic characterization of real numbers.)

4.3 Kochen and Specker Theorem

While it is quite easy to conceive of theoretical examples of classical tautologies, which are not quantum tautologies – the distributivity rule is such an example – it is much harder to relate such propositions to real measurable properties of physical systems. This, among other things, is achieved by an elegant result of Kochen and Specker (1967).

As before let $\{a_1, a_2, ..., a_n, ...\}$ be a set of atomic propositions. Define

$$b(a_i, a_j) = [\sim(a_i \wedge a_j)]$$

$$c(a_i, a_j, a_k) = (a_i \vee a_j \vee a_k)$$

(4-2)

Classically $b(a_i, a_j)$ is true if and only if at least one of the propositions a_i, a_j is false. Quantum logically $b(a_i, a_j)$ is certainly true in these classical cases but it *may* also be true in some cases when both a_i and a_j are true. Similarily $c(a_i, a_j, a_k)$ is classically true if and only if at least one of the propositions a_i, a_j, a_k is true. Quantum logically $c(a_i, a_j, a_k)$ is certainly true in all classical cases but it may also be true even when a_i, a_j, a_k are all false. Now consider a proposition with 10 atoms:

$$d = d(a_1, a_2, ..., a_{10}) = b(a_1, a_2) \wedge b(a_1, a_3) \wedge b(a_1, a_9) \wedge b(a_2, a_4)$$

$$\wedge b(a_2, a_6) \wedge b(a_3, a_5) \wedge b(a_3, a_7) \wedge b(a_4, a_6) \wedge b(a_4, a_8) \wedge b(a_5, a_7)$$

$$\wedge b(a_5, a_8) \wedge b(a_6, a_7) \wedge b(a_8, a_9) \wedge b(a_8, a_{10}) \wedge b(a_9, a_{10})$$

$$\wedge c(a_2, a_4, a_6) \wedge c(a_3, a_5, a_7) \wedge c(a_8, a_9, a_{10})$$

(4-3)

The best way to understand the meaning of that proposition is by considering the following graph Fig. (4-1) call it G:

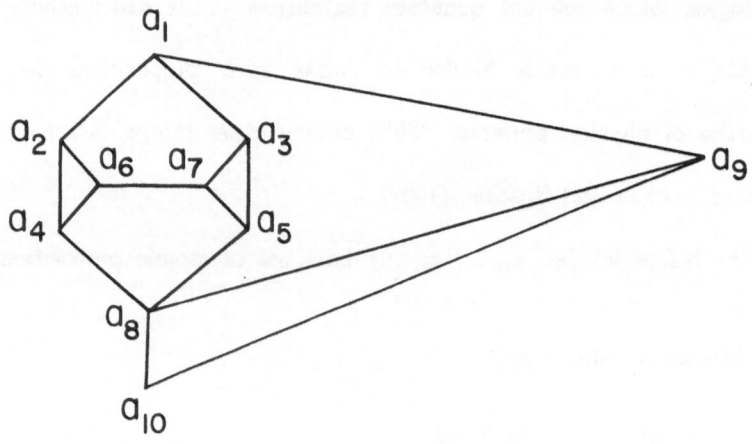

Fig. (4-1): The Graph G

The vertices of G are just our proposition symbols a_1, a_2, ..., a_{10}. Two propositions a_i, a_j, $1 \leq i < j \leq 10$ are connected by an edge in G if and only if $b(a_i, a_j)$ appears as a conjunct in the proposition d. Moreover the proposition $c(a_i, a_j, a_k)$, $1 \leq i < j < k \leq 10$ appears as a conjunct in d if and only if all three pairs

$\{a_i, a_j\}$, $\{a_i, a_k\}$, $\{a_j, a_k\}$ are edges in G, that is, if and only if $\{a_i, a_j, a_k\}$ is a triangle in G.

Let $t:\{a_1, a_2, ..., a_{10}\} \rightarrow \{0,1\}$ be a *classical* truth function such that $t(d) = 1$ then no two atomic propositions connected by an edge in G can have the truth value 1, otherwise $t(a_i) = t(a_j) = 1$ entail $t(b(a_i,a_j)) = 0$ hence $t(d) = 0$.

Lemma (4-4): If $t:\{a_1, ..., a_{10}\} \rightarrow \{0,1\}$ is a classical truth function such that $t(d) = 1$ and $t(a_1) = 1$ then $t(a_{10}) = 1$

Proof: Suppose by contrast that $t(a_1) = t(d) = 1$ but $t(a_{10}) = 0$. Now $t(a_1) = 1$ entails $t(a_2) = t(a_3) = t(a_9) = 0$, we know that $t(a_9) = t(a_{10}) = 0$, hence since $t(c(a_8, a_9, a_{10})) = 1$ it follows that $t(a_8) = 1$. Therefore $t(a_4) = t(a_5) = 0$, also $t(a_2) = t(a_3) = 0$ $t(c(a_2, a_4, a_6)) = 1$, $t(c(a_3, a_5, a_7)) = 1$ and thus we conclude that $t(a_6) = t(a_7) = 1$. But this is a contradiction since $t(b(a_6, a_7)) = 1$.

Now consider the graph Γ given in Fig. (4-2):

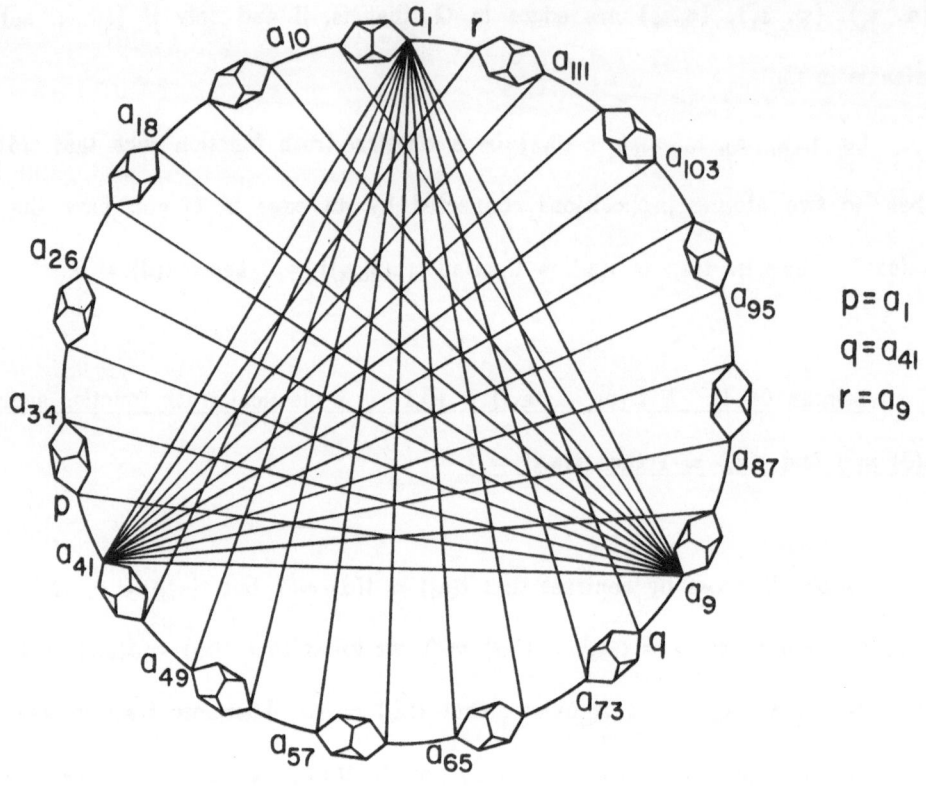

Fig. (4-2): The graph Γ

It consists of 117 vertices a_1, a_2, ..., a_{117}. (Note that we identify $p = a_1$, $q = a_{41}$, $r = a_9$.) Let $e = e(a_1, a_2, ..., a_{117})$ be the proposition which is a connjunction of all the $b(a_i, a_j)$ such that $\{a_i, a_j\}$ is an edge in Γ and $c(a_i, a_j, a_k)$ such that $\{a_i, a_j, a_k\}$ is a traingle in Γ

$$e = \bigwedge_{\{a_i, a_j\} \text{ edge in } \Gamma} b(a_i, a_j) \quad \wedge \quad \bigwedge_{\{a_i, a_j, a_k\} \text{ triangle in } \Gamma} c(a_i, a_j, a_k) \qquad (4\text{-}4)$$

e is a classical logical falsity that is, t(e) = 0 for all classical truth functions

t:{a_1, a_2, ..., a_{117}} → {0,1}. For suppose that t is such a classical truth function

and suppose that t(e) = 1 then t(c(a_1, a_9, a_{41})) = 1 since a_1, a_9, a_{41} is a triangle

in Γ. Therefore at least one of the atomic propositions a_1, a_9, a_{41} is true. Suppose

t(a_1) = 1 then by Lemma (4-4) t(a_{10}) = 1 and thus by the same lemma

t(a_{18}) = 1 and so on . . . t(a_{41}) = 1 and thus t(a_{41}) = 1 t(b(a_1, a_{42})) = 0. But

{a_1, a_{41}} is an edge in Γ hence we conclude t(e) = 0, contradiction. In the same

way we can derive a contradiction from the assumption t(a_9) = 1 or t(a_{41}) = 1.

So far we have constructed a very complex classical logical falsity e. We

shall show that in some circumstances e can have quantum truth value 1. More

specifically: *One can substitute instead of the a_1-s actual propositions which refer to*

a well defined state description of a concrete physical system and obtain a case in

which e is true.

To see this consider a spin-1 massive particle or equivlently a system whose

principal quantum number is n=1 and azymutal quantum number is j=2. Let x,y,z

be three orthogonal directions in space. The angular momentum operators for the

above system are represented by:

$$J_x = \frac{1}{\sqrt{2}} \begin{pmatrix} 0 & 1 & 0 \\ 1 & 0 & 1 \\ 0 & 1 & 0 \end{pmatrix} \qquad J_y = \frac{1}{\sqrt{2}} \begin{pmatrix} 0 & -i & 0 \\ i & 0 & -i \\ 0 & i & 0 \end{pmatrix} \qquad J_z = \begin{pmatrix} 1 & 0 & 0 \\ 0 & 0 & 0 \\ 0 & 0 & -1 \end{pmatrix}$$

We have $J_x^2 + J_y^2 + J_z^2 = 2I$ (I is the unit matrix). It turns out that *in this*

particular representation J_x^2, J_y^2, J_z^2 *pairwise commune* (this does not happen in

irreducible representations in higher dimensions). So for this representation

$[J_x^2, J_y^2] = [J_x^2, J_s^2] = [J_y^2, J_s^2] = 0$. This means, theoretically that a simultaneous value for J_x^2, J_y^2, J_s^2 can be measured. (Indeed Kochen and Specker provide a perscription for such a measurement on an orthohelim atom in the 2^3S_1 state.) The eigenvalues of J_x^2, J_y^2, J_s^2 are either zero or one and $J_x^2 + J_y^2 + J_s^2 = 2I$. We conclude: *in one and only one out of three orthogonal directions x,y,z the (square of the) angular momentum is zero.*

This statement is valid for every orthogonal tripple x,y,z (although it cannot be verified simultaneously for two or more distinct orthogonal tripples). In particular there could exist no orthogonal pair x,y such that the angular momentum is zero in both the x and the y directions simultaneously.

The connection between this physical system and the proposition e is given by the following:

Lemma (4-5): There exists a one one map $a_i \to x_i$ $1 \leq i \leq 117$ between the set of atomic propositions $\{a_1, a_2, ..., a_{117}\}$ and a set of 117 unit vectors in three dimensional real space such that $\{a_i, a_j\}$ is an edge in the graph Γ if and only if x_i and x_j are orthogonal.

Before proving the lemma let us see how to apply it. Given the directions x_1, x_2, ..., x_{117} consider the 117 propositions s_i – "the angular momentum in the x_i direction is zero." Since we are dealing with a system for which the angular momentum cannot be zero in two orthogonal directions, the proposition $b(s_i, s_j)$ is true for all i,j such that $\{a_i, a_j\}$ is an edge in the graph Γ. Moreover, since in our system the angular momentum is zero in one (and only one) out of every three orthogonal directions, we conclude that $c(s_i, s_j, s_k)$ is true whenever $\{a_i, a_j, a_k\}$ is a triangle in Γ. It follows that

$$e(s_1, s_2, ..., s_{117}) = \bigwedge_{\{a_i, a_j\} \text{ edge in } \Gamma} b(s_i, s_j) \wedge \bigwedge_{\{a_i, a_j, a_k\} \text{ triangle in } \Gamma} c(s_i, s_j, s_k)$$

is true, in spite of the fact that e is a classical logical falsity! In the next sections we shall see how this surprising result can be interpreted (surely its interpretation depends on the meaning we assign to the quantum truth functions). From a purely formal standpoint, however, the result can easily be understood. We may have a quantum truth function θ such that $\theta(a_i) = \theta(a_j) = 1$ and also $\theta(b(a_i, a_j)) = 1$ or such that $\theta(a_i) = \theta(a_j) = \theta(a_k) = 0$ but $\theta(c(a_i, a_j, a_k)) = 1$. Hence e is surely not a *quantum* logical falsity.

We conclude this section by proving the last result.

Proof of Lemma (4-5): Consider first the graph G in fig. (4-1) we shall prove:

There are ten unit vectors in $R^{(3)}$; $x_1, x_2, ..., x_{10}$ such that the angle between x_1 and x_{10} is $\pi/10$ and such that for $1 \leq i < j \leq 10$ $\{a_i, a_j\}$ is an edge in the graph G if and only if x_i is orthnogonal to x_j.

To see this let x_{10} be an arbitrary unit vector. Choose any ortnogonal tripple of unit vectors x,y,z and let ξ, η be two real numbers whose values will be fixed shortly.

Let $x_6 = x$, $x_7 = z$, $x_2 = (y + \xi z)(1 + \xi^2)^{-1/2}$, $x_3 = (x + \eta z)(1 + \eta^2)^{-1/2}$.

Then x_4 is fixed , up to a $\pm$ sign, by the requirement that x_2, x_4, x_6 is an orthogonal tripple and similarily x_5 is fixed by the requirements that x_3, x_5, x_7 is an orthogonal tripple. That is, $x_4 = (\xi y - z)(1 + \xi^2)^{-1/2}$,

$x_5 = (\eta x - y)(1 + \eta^2)^{-1/2}$. The condition that x_1 should be orthogonal to both x_2 and x_3, and that x_8 be orthogonal to both x_4 and x_5 give the values:

$$x_1 = (\xi \eta x - \xi y + z)(1 + \xi^2 + \eta^2 \xi^2)^{-1/2}$$

$$x_8 = (x + \eta y - \xi \eta z)(1 + \eta^2 + \eta^2 \xi^2)^{-1/2}$$

x_9 is chosen by the requirement that it is orthogonal to both x_1 and x_{10}. Since x_8 is orthogonal to x_9 we conclude that x_8 lies in the plane spanned by x_1 and x_{10}, also x_8 is orthogonal to x_{10}. Since the angle between x_1 and x_{10} should be

$\dfrac{\pi}{10}$,the angle between x_1 and x_8 is $\dfrac{\pi}{2} - \dfrac{\pi}{10}$. Hence we should have:

$$\cos(\frac{\pi}{2} - \frac{\pi}{10}) = \sin(\frac{\pi}{10}) = \frac{\xi \eta}{[(1 + \xi^2 + \xi^2 \eta^2)(1 + \eta^2 + \xi^2 \eta^2)]^{1/2}}$$

And it is easy to see that a choice of two numbers ξ, η which satisfy this requirement is possible. Having made that choice we proved our claim. To complete the construction of the map from the larger graph Γ in Fig. (4–2), fix ten directions x_1, ..., x_{10} which represent the graph G and such that the angle between

x_1 and x_{10} is $\dfrac{\pi}{10}$. Now rotate the system of 10 vectors by $\dfrac{\pi}{10}$ around the x_9 axis.

Call this rotation α, then $\alpha(x_1) = x_{10}$, $\alpha(x_9) = x_9$, and the angle between x_{10} and

$\alpha(x_{10})$ is $\dfrac{\pi}{10}$. The vectors $\alpha(x_1)$, ..., $\alpha(x_{1o})$ exhibit the same orthogonality relations as x_1, ..., x_{10} respectively. Put $x_{11} = \alpha(x_2)$, $x_{12} = \alpha(x_3)$ ... and so on (remember that $\alpha(x_9) = x_9$). Repeat the rotation 5 times then $\alpha^5(x_1) = x_{41}$ is orthogonal to x_1. Now rotate the entire picture once by $\dfrac{\pi}{2}$ around the x_1 axis and once by $\dfrac{\pi}{2}$ around the x_{41} axis, the resulting 117 vectors satisfy the requirements of Lemma (4–5). ∎

4.4   Realistic Quantum Logic I: Conspiratorial Interpretation

By "realistic quantum logic" I shall refer to those interpretations of the quantum truth functions which take them to represent actual state of affairs in the microphysical world.   In this section I shall introduce such an interpretation, "the conspiratorial approach" which is really more a satire than a serious proposal, but which can nevertheless teach us some important lessons.

Given two propositions b_1, b_2 quantum logic postulates five states of affairs with respect to the truth value of the conjunction $b_1 \wedge b_2$.   These are given in the following truth table:

$\theta(b_1)$	$\theta(b_2)$	$\theta(b_1 \wedge b_2)$
0	0	0
0	1	0
1	0	0
1	1	1
1	1	0

The first four are just the classical truth functions while the fifth is the peculiar one. All four classical truth functions are going to be interpreted in the usual manner. The fifth possibility reads as follows:

$\theta(b_1) = \theta(b_2)$ and $\theta(b_1 \wedge b_2) = 0$ *means that in reality both* b_1 *and* b_2 *are true and of course so is* $b_1 \wedge b_2$. *But this state of affairs is hidden from us. When we check experimentally whether* b_1 *(and only* b_1*) is true we find that it is. When we check whether* b_2 *(only* b_2*) is true we find that it is. But if we check directly whether the conjunction* $b_1 \wedge b_2$ *is true we shall be mislead to believe that it is false. In this case our equipment will wrongly report that either* b_1 *is false (with certain probability) or that* b_2 *is false (again with certain probability).*

The quantum truth value of a disjunction $\theta(b_1) = 0$, $\theta(b_2) = 0$, $\theta(b_1 \vee b_2) = 1$ should be interpreted in a similar way. The conspiratorial view, we see, resurrects the long sleeping Cartesian demon. When we attempt to know too much (conjunction), or too little (disjunction), the Cartesian demon sometimes misleads us to believe in false propositions ("sometimes", because afterall the classical truth functions are also quantum truth functions).

Let us take the Kochen and Specker proposition, from the previous section, as an example. There we had

$$
e = \bigwedge_{\{a_i,a_j\} \text{ edge in } \Gamma} [\sim(s_i \wedge s_j)] \wedge \bigwedge_{\{a_i,a_j,a_k\} \text{ triangle in } \Gamma} (s_i \vee s_j \vee s_k)
$$

where s_i is the proposition "the spin is zero in the x_i- direction". $\sim(s_i \wedge s_j)$ states that the spin is not zero in both the x_i and x_j direction and $(s_i \vee s_j \vee s_k)$ states that the spin is zero in at least one of the directions x_i, x_j, x_k. We have seen that e is quantum mechanically true, in spite of the fact that e is a classical logical falsity.

The fact that e is true when we substitute the propositions s_i means that in that case $\theta(e) = 1$ for some quantum truth function, hence $\theta(\sim(s_i \wedge s_j)) = 1$ and $\theta(s_i \vee s_j \vee s_k) = 1$, whenever $\{a_i, a_j\}$ is an edge in Γ, and $\{a_i, a_j, a_k\}$ a triangle in Γ. But θ cannot be a classical truth function hence either $\theta(s_i) = \theta(s_j) = 1$ and $\theta[\sim(s_i \wedge s_j)] = 1$ for some edge $\{a_i, a_j\}$ of Γ or $\theta(s_i) = \theta(s_j) = \theta(s_k) = 0$ and $\theta(s_i \vee s_j \vee s_k) = 1$ for some triangle $\{a_i,a_j,a_k\}$ in Γ. We shall take each case in turn.

According to the conspiratorial view $\theta(s_i) = \theta(s_j) = 1$ and $\theta(s_i \wedge s_j) = 0$ (which implies $\theta[\sim(s_i \wedge s_j)] = 1$) means that when we measure the spin in the x_i-direction *alone* we shall find it to be zero, when we measure the spin in the x_j direction *alone* we shall find it to be zero but when we measure the spin (or rather the square of the spin) in both directions *at once* – and this is possible since in our case $J_{x_i}^2$, $J_{x_j}^2$ commune – we shall find that at least one of the spins is non zero. Similarly $\theta(s_i) = \theta(s_j) = \theta(s_k) = 0$ but $\theta(s_i \vee s_j \vee s_k) = 1$ means that a measurement of the spin in the x_i direction alone or the spin in the x_j direction alone, or the spin in the x_k direction alone will show, in each case, that spin $\neq 0$. But if we measure the (square of the) spins in all three directions x_i, x_j, x_k simultaneously we shall find that one of the values is zero. (This is not a completely nonsensical view, since the experiment involved in a measurement of $J_{x_i}^2$ alone, is different from the experiment involved in the simulatneous measurement of $J_{x_i}^2$, $J_{x_j}^2$, $J_{x_k}^2$).

Since the conspiratorial interpretation is realistic, it means that "in reality" the rule *"the spin value is zero in one and only one out of every three orthogonal directions"* is false. There are orthogonal tripples for which the spin is non zero in all three directions or such that the spin is zero in more than one direction. But we shall never be able to detect such tripples because, when we try, our equipment will mislead us to believe that the rule has not been broken.

The conspiratorial interpretation is consistent with quantum statistics, the reason is that quantum correlations can always be represented as weighted averages of the quantum truth values. The weights can be so chosen as to yield the correct statistical result. The trouble with the conspiratorial view (as with all realistic interpretations of quantum logic) lies with special relativity and not with quantum mechanics. The "state of affairs" described by the conspiratorial interpretation of quantum logic are non covariant. To see that we shall return to the Clauser–Horne version of the E.P.R. experiment described in Section 3.7. We were considering electron pairs in the singlet state. In the language of propositions, the projections (3–12) describe four possible states of affairs:

s_1 – the left electron in the pair has spin up in the x direction

s_2 – the left electron in the pair has spin up in the y direction

s_3 – the right electron in the pair has spin up in the z direction

s_4 – the right electron in the pair has spin up in the w direction

We shall take the particular case in which x,y,w are coplanner forming angles of $120°$ and in which y = z. Then for n = 4, S = {{1,3}, {1,4}, {2,3}, {2,4}} we have $p = (p_1, p_2, p_3, p_4, p_{13}, p_{14}, p_{23}, p_{24}) = (\frac{1}{2}, \frac{1}{2}, \frac{1}{2}, \frac{1}{2}, \frac{3}{8}, \frac{3}{8}, 0, \frac{3}{8}) \notin c(4,S)$.

Since $p \notin c(4,S)$, we cannot represent p as a convex combination of the classical truth functions u^ε, $\varepsilon \in \{0,1\}^4$ alone. But we can always represent p as a convex combination of the vertices of q(4,S) that is, the quantum truth functions. In any such representation at least one of the non–classical truth functions will appear with a non–zero coefficient.

Consider for example one such truth function:

$u = (u_1, u_2, u_3, u_4, u_{13}, u_{14}, u_{23}, u_{24}) = (1,0,1,0,0,0,0,0)$. This is a case of quantum truth function θ such that $\theta(s_1) = \theta(s_3) = 1$, $\theta(s_2) = \theta(s_4) = 0$, $\theta(s_1 \wedge s_3) = \theta(s_1 \wedge s_4) = \theta(s_2 \wedge s_3) = \theta(s_2 \wedge s_4) = 0$. The non classical component of θ is: $\theta(s_1) = \theta(s_3) = 1$, $\theta(s_1 \wedge s_3) = 0$.

According to the conspiratorial interpretation, a particle pair whose state is described by the truth function θ has the following properties: If we put a Stern–Gerlach magnet oriented in the x direction to the left of the source and no magnet to the right of the source then the left electron will definitely go "up", since s_1 is true. If we put a Stern–Gerlach magnet oriented in the z direction to the right of the source, and no magnet to the left of the source the right electron will go up, since s_3 is true. But if we put two Stern–Gerlach magnets, one oriented in the x–direction to the left of the source, and another oriented in the z–direction to the right of the source, then either the left electron will go down (with 50% chance, say) or the right electron will go down (with 50% chance). All this happens since $s_1 \wedge s_2$ is false, that is our equipment will fail to detect the actual state of affairs.

Note that in order to make our interpretation consistent we have to take a similar stand with respect to disjunctions. Consider the truth function
$u' = (u'_1, u'_2, u'_3, u'_4, u'_{13}, u'_{14}, u'_{23}, u'_{24}) = (0,1,0,1,0,1,0,0)$ (note this is a classical truth function for a conjunction). We have $\theta'(s_1) = \theta'(s_3) = 0$ and surely $\theta'(s_4 \wedge s_3) = 0$ but we *may* have $\theta'(s_1 \vee s_3) = 1$, hence by definition: $\theta'[(\sim s_1) \wedge (\sim s_2)] = 0$. This means that for an electron pair whose state is described by θ' we have the following: If we put only one of the magnets either the one oriented in the x–direction to the left of the source, or the one oriented in the y–direction to the right of the source (but not both), then the respective electrons will go down. But if we put both magnets then, since $\theta'[(\sim s_1) \wedge (\sim s_3)] = 0$, either the left electron will go "up" or the right electron will go "up".

In order to see how the entire experiment can be explained in this way
consider for example the representation

$$p = (p_1, p_2, p_3, p_4, p_{13}, p_{14}, p_{23}, p_{24}) = (\frac{1}{2}, \frac{1}{2}, \frac{1}{2}, \frac{1}{2}, \frac{3}{8}, \frac{3}{8}, 0, \frac{3}{8})$$

$$= \frac{1}{2}(0,0,0,0,0,0,0,0) + \frac{3}{8}(1,1,1,1,1,1,0,1) + \frac{1}{8}(1,1,1,1,0,0,0,0)$$

The "conspiratorial" meaning of each one of the three truth functions in this
convex combination is given by the following table, which indicates what happens
when we put *two* magnets one on the left and one on the right:

State	magnets orientation		observed results				frequency
	left	right	x	y	z	w	
(0,0,0,0,0,0,0,0)	x	z	up	–	down	–	1/8
			down	–	up	–	1/8
			down	–	down	–	3/4
	x	w	up	–	–	down	1/8
			down	–	–	up	1/8
			down	–	–	up	3/4
	y	z	–	up	down	–	1/2
			–	down	up	–	1/2
	y	w	–	up	–	down	1/8
			–	down	–	up	1/8
			–	down	–	down	3/4
(1,1,1,1,1,1,0,1)	x	z	up	–	up	–	1
	x	w	up	–	up	–	1
	y	z	–	up	down	–	1/2
			–	down	up	–	1/2
	y	w	–	up	–	up	1
(1,1,1,1,0,0,0,0)	x	z	up	–	down	–	1/2
			down	–	up	–	1/2
	x	w	up	–	–	down	1/2
			down	–	–	up	1/2
	y	z	–	up	down	–	1/2
			–	down	up	–	1/2
	y	w	–	up	–	down	1/2
			–	down	–	up	1/2

take for example the first case in the table. We have an electron pair for which the truth value is $\theta(s_1) = \theta(s_2) = \theta(s_3) = \theta(s_4) = 0$ and then surely all the conjunctions are false. The disjunction $s_1 \vee s_3$ can however take two truth values $\theta(s_1 \vee s_3) = 0$, which is the classical case, and $\theta(s_1 \vee s_3) = 1$. $\theta(s_1 \vee s_3) = 0$ means that $\theta((\sim s_1) \wedge (\sim s_3)) = 1$ hence if we put magnets, on the x direction on the left, and the z direction on the right, both electrons will go down. According to the table this case occures 3/4 of the times. $\theta(s_1 \vee s_3) = 1$, means $\theta((\sim s_1) \wedge (\sim s_3)) = 0$, hence either both electrons go "up", this however never occurs, according to the table, or the left goes "up", the right go "down" (1/8 of the time) or the left go "down" the right go "up" (1/8 of the time). Note that if we put only one magnet either the left one of the right one (but not both) the respective electrons will definitely go down when the state of affairs is given by θ.

In this way all the cases of the table should be interpreted. Thus in the case the left magnet is in the x direction, and the right magnet is in the z direction, the frequency of electrons that go up on the left is: $\frac{1}{2} \times \frac{1}{8} + \frac{3}{8} \times 1 + \frac{1}{8} \times \frac{1}{2} = \frac{1}{2}$ and the frequency of electrons that go up in both directions is $\frac{3}{8} \times 1 = \frac{3}{8}$. Similarly when we put the left magnet in the y–direction and the right magnet in the z direction the rate of electrons that go up on the left is $\frac{1}{2} \times \frac{1}{2} + \frac{1}{2} \times \frac{3}{8} + \frac{1}{2} \times \frac{1}{8} = \frac{1}{2}$ and the rate of electrons that go down in both sides is 0. It is therefore evident that the conspiratiorial interpretation is consistent with the frequencies of qunatum mechanics. We shall see now that this interpretation is utterly non local.

Consider for example the truth function u = (1,1,1,1,0,0,0,0). According to our analysis the state of 1/8 of the electron pairs is described by u. Suppose for the moment that we have some way to control the source and isolate only those pairs which are in this particular state, namely the state described by the truth function u. Then we can transmit messages superluminally with high probablity. The argument here is identical to our analysis of the hidden varible theories in Section (3.11): If we put only one magnet to the left of the source, oriented in the x direction, the left electron will go "up". If we introduce a magnet oriented in the z direction on the right, then with 50% change the *left* electron will go down. Thus we can transmit a message from right to left which, after N electrons have arrived on the left, will be delivered with probability $1-2^{-N}$ that is, almost certainty. Since the distance between the magnets is arbitrary, and since the flux of electrons can be made large, the message is being delivered superluminally almost surely. Again the conspiratorial interpretation of quantum logic does not necessarily imply that we can control the "truth values" and isolate those electron pairs which are in a state descirbed by u. What we have demonstrated though is that this state of affairs is non covariant. The conspiratorial view therefore, turns the theory of relativity into a statistical, rather than principal theory.

Of course the same analysis can be carried out with respect to any non classical truth function u. The similarity between the conspiratorial view and hidden variable theories is no accident. The conspiratorial view is, in fact, a hidden variable theory in disguise.

Given a physical system S we can identify the property P with the set of all quantum truth functions which make the statement "S has property P" true, call this set A(P). Then for two properties P and Q we may have:

A(P ∧ Q) ≠ A(P) ∩ A(Q), since in quantum logic the statements "S has property P" and "S has property Q" may each be true but their conjunction false. This is precisely what leads to the violation of locality indicated in Section (3.11).

As far as I know, nobody ever seriously proposed the conspiratorial interpretation. But views which are essentially identical with it have been proposed. We shall consider one in the following section.

4.5   Realistic Quantum Logic II: The Operational Interpretation

This seemingly more radical proposal was introduced by D. Finkelstein (1962, 1968) and became a subject of philosophical discussion when it was adopted by H. Putnam (1968) as a clarification and interpretation of Birkhoff and von Neumann (1943) original claims.

Ideally, with every property P of a physical system we can associate a test T such that the system has the property P if and only if it "passes" the test T, more precisely, *it would pass the test T, if T were performed.*

Now suppose that T_1, T_2 are two tests such that every system which "passes" T_1 also "passes" T_2, in that case we put $T_1 \leq T_2$. If R_1, R_2 are the properties associated with the test T_1, T_2 respectively then we can also put $R_1 \to R_2$, that is having property R_1 implies having property R_2. In this way we obtain an operational definition of the implication relation. Putnam justly stresses that the term "operational definition" should be taken with a grain of salt. The idealized tests T_1, T_2 – like the idealized rods and clocks of special relativity – are themselves theoretical objects.

Since we seek an interpretation of the logical connectives, no harm is implied by such idealizations, provided that we remember that the idealized tests are only approximations of real experiments. Consequently the "operational definitions" do not, strictly speaking, define the logical connectives, they only indicate how they should be used in practical physical discourse.

Having "defined" the implication relation $\leq$ we can proceed and "define" the other logical connectives in the same manner. The conjunction $T_1 \wedge T_2$ is the greatest lower bound, (if it exists) on all tests T, such that everything which "passes" T also passes T_1 and passes T_2.

Now suppose that we apply these "definitions" to qunatum mechanical properties. Quantum properties typically have the form: "The observable A has value a," and every such property is associated with a closed subspace of a Hilbert space, or equivalently with the projection on that subspace. With this identification it is not difficult to see (see e.g. Finkelstein 1968) that $T_1 \leq T_2$ corresponds to subspace inclusion, and $T_1 \wedge T_2$ corresponds to subspace intersection.

The difference between quantum logic, so conceived and the more traditional interpretation of Bohr (which Putnam calls "Copenhagen double think"), is in its adoption of "counterfactual definiteness". We can assign a property to a system independently of a measurement. The system has property R not just in case it actually passes the relevant test, but also in case we know it *would* pass the test *if* it were performed. Thus Finkelstein maintains:

"In the two slit experiment did the electron reaching the screen have to go either through one slit or through another? Common sense says yes; we say yes; Bohr says (I think) that the question is meaningless."

But the ascription of properties to particles independently of an actual measurement leads to difficulties of the kind we have already encountered a few times. Suppose that ψ is a normalized vector in a Hilbert space. We can find two properties, associated with the projections E_1, E_2 such that

$$p_1 = <\psi, E_1\psi> = 0.9, \quad p_2 = <\psi, E_2\psi> = 0.9 \quad \text{and}$$

$$p_{12} = <\psi, (E_1 \wedge E_2)\psi> = 0$$

Let T_1, T_2 be the tests associated with the projections E_1, E_2 respectively and consider a source of particles all in the state $W = E_\psi$. Then 90 out of every 100 particles would pass the test T_1 *if* that test were performed. Ninety out of every 100 particles would pass test T_2 *if* that test were performed, and no particle would pass $T_1 \wedge T_2$ if this test were performed. Hence, we must admit that among the particles in the source there are those for which the statement: "The particle has the property E_1" is "operationally" true, and the statement "The particle has property E_2" is "operationally" true, but the conjunction "The particle has both property E_1 and property E_2" is "operationally" false. We are back with the quantum truth function $\theta(a_1) = \theta(a_2) = 1$, $\theta(a_1 \wedge a_2) = 0$. Hence we conclude: *There is no "operational" distincition between the conspiratorial interpretation and the operational interpretation.* Of course there is a "metaphysical" difference between these two approaches. The operational view does not assume the existence of a "Cartesian demon" who mislead us to believe in false statements. Operationalism is more rational but not less conservative. Like the conspiratorial view it is a hidden variable theory in disguise. In this context Putman's dictum: "Don't seek to have a revolution and minimize it too" has an ironic ring to it, afterall, this is precisely what the operational interpretation seeks to achieve.

What is the source of trouble? I have stressed many times that associating actual state of affairs with quantum "truth functions" is in fact an *extention* of quantum mechanics. Formally this follows from theorem (3–8): The non classical vertices of $q(n,S)$ – the "genuine" quantum truth functions – are not themselves elements of $q(n,S)$. They do not even represent possible quantum frequencies. If we assign reality to these truth functions we play a different game.

No wonder that the operational view is non local, in the same sense that the conspiratorial view of the previous section, and the hidden variable approach of Section 3.11 are non local. If we adopt the operational interpretation we must admit that there exist pairs of electrons in the singlet state for which the following three statements are simultaneously true:

"If we put a Stern–Gerlach magnet, oriented in the x direction to the left of the source, and no magnet to the right of the source, the left electron would go up".

"If we put a Stern–Gerlach magnet, oriented in the z direction to the right of the source, and no magnet to the left of the source, the right electron would go up".

"If we put two Stern Gerlach magnets, one oriented in the x direction to the left of the source, and one oriented in the z direction to the right of the source, then either the left electron would go down or the right electron would go down".

This follows from the fact that the frequencies observed in the Clauser Horne version of the E.P.R. experiment violate the facet inequalities of c(n,S). The details of the analysis are precisely the same as in the conspiratorial interpretation of the previous sections. We have already seen that the very existence of such electron pairs, regardless of whether we can isolate them or not, contradicts the principle of locality and thus poses a problem for the operational view.

4.6 Realistic Quantum Logic and the Entropy Principle

Since realistic quantum logic represents an extension of quantum mechanics we may expect it to lead to some novel predictions. The violation of locality is one such prediction, which we may, nevertheless, not be able to test directly. Let us ignore this difficulty for the moment, and seek a way to obtain more definitive results.

Suppose that the quntum truth values represent some actual state of affairs, either in accord with the conspiratiorial view, or the operational view or some other interpretation. In this framework we face a problem similar to the one we faced in the classical case (Section 2.8). Given $p \in q(n,S)$ there is usually more than one way to represent p as a convex combination of the vertices of $q(n,S)$, the quantum truth functions. Here again we may use the entropy principle. Thus if $p \in q(n,S)$ we shall seek a representation

$$p = \sum_{k=1}^{r} \lambda_k u^k$$

where u^1, u^2, ..., u^r are all the vertices of $q(n,S)$, and the following is satisfied:

$$\lambda_k \geq 0 \qquad k = 1, 2, ..., r$$

$$\sum_{k=1}^{r} \lambda_k = 1$$

$$\sum_{k=1}^{r} \lambda_k u_i^k = p_i \qquad i = i, 2, ..., n$$

$$\sum_{k=1}^{r} \lambda_k u_{ij}^k = p_{ij} \qquad \{i,j\} \in S$$

and such that

$$H = -\sum_{k=1}^{r} \lambda_k lg\lambda_k$$

is maximal.

We are justified in looking for such a representation if we have good reasons to believe that, apart from the correlations at hand, the system is maximally mixed with respect to the truth values. In the quantum mechanical case we do not, in general, have that kind of information. Hence we shall adopt the principle of entropy without further ado.

This principle, which makes sense only for a realist quantum logic, allows us to "predict" results which lie beyond the scope of quantum mechanics. Take for example the Bell Wigner version of the E.P.R. experiment (Section 3.8). There we had n = 3, S = {{1,2}, {1,3}, {2,3}} and p ∈ q(n,S) was given by

$p = (p_1, p_2, p_3, p_{12}, p_{13}, p_{23}) = (\frac{1}{2}, \frac{1}{2}, \frac{1}{2}, \frac{1}{8}, \frac{1}{8}, \frac{1}{8})$. Now suppose that we could

check experimentally how many particle pairs have the following property: "The left particle has spin up in both the x and y directions and the right particle has spin down in both y and z directions." Quantum mechanics provides no account of such experiments, according to Bohr the conjunction of these properties is meaningless. Realist quantum logic together with the entropy principle provides a definite answer: The frequency of pairs with the above property is just the coefficient of $u = (u_1,u_2,u_3,u_{12},u_{13},u_{23}) = (1,1,1,1,1,1)$ in the convex combination

$\Sigma \lambda_k u^k = p = (\frac{1}{2}, \frac{1}{2}, \frac{1}{2}, \frac{1}{8}, \frac{1}{8}, \frac{1}{8})$ for which $-\Sigma \lambda_k \lg \lambda_k$ is maximal.

Indeed $u = (1,1,1,1,1,1)$ is the only quantum truth value for which all the above properties hold and for which we shall be "able" to *detect* them (as opposed to $(1,1,1,1,0,1)$ for example, in which our supposed equipment will fail to detect the true state of affairs). $q(3,S)$ has 18 vertices and the most general representation of p is thus:

$$p = (\frac{1}{2}, \frac{1}{2}, \frac{1}{2}, \frac{1}{8}, \frac{1}{8}, \frac{1}{8}) = \lambda_1(0,0,0,0,0,0) + \lambda_{21}(1,0,0,0,0,0) + \lambda_{22}(0,1,0,0,0,0) +$$

$$\lambda_{23}(0,0,1,0,0,0) + \lambda_{31}(1,1,0,1,0,0) + \lambda_{32}(1,0,1,0,1,0) + \lambda_{33}(0,1,1,0,0,1) +$$

$$\lambda_{41}(1,1,0,0,0,0) + \lambda_{42}(1,0,1,0,0,0) + \lambda_{43}(0,1,1,0,0,0) + \lambda_5(1,1,1,0,0,0) +$$

$$\lambda_{61}(1,1,1,0,0,0) + \lambda_{62}(1,1,1,0,1,0) + \lambda_{63}(1,1,1,0,0,1) + \lambda_{71}(1,1,1,1,1,0) +$$

$$\lambda_{72}(1,1,1,0,0,0) + \lambda_{73}(1,1,1,1,0,1) + \lambda_8(1,1,1,1,1,1)$$

From symmetry consideration it is obvious that for maximal entropy we should take $\lambda_{k1} = \lambda_{k2} = \lambda_{k3}$ for $k = 2,3,4,6,7$. So we have the equations:

$$\lambda_3 + \lambda_6 + 2\lambda_7 + \lambda_8 = \frac{1}{8}$$

$$\lambda_2 + 2\lambda_3 + 2\lambda_4 + \lambda_5 + 3\lambda_6 + 3\lambda_7 + \lambda_8 = \frac{1}{2} \qquad (4\text{-}5)$$

$$\lambda_1 + 3\lambda_2 + 3\lambda_3 + 3\lambda_4 + \lambda_5 + 3\lambda_6 + 3\lambda_7 + \lambda_8 = 1$$

We solve these equations and represent λ_1, λ_5, λ_8 as linear combinations of λ_2, λ_3, λ_4, λ_6, λ_7 next we put:

$$0 = -\frac{\partial H}{\partial \lambda_j} = \lg(e\lambda_1)\frac{\partial \lambda_1}{\partial \lambda_j} + \lg(e\lambda_5)\frac{\partial \lambda_5}{\partial \lambda_j} + \lg(e\lambda_8)\frac{\partial \lambda_8}{\partial \lambda_j} + 3\,\lg(e\lambda_j)$$

for $j = 2,3,4,6,7$ where H is the entropy. Substituting we obtain

$$\lambda_2^3 = \lambda_1^2\lambda_5 \quad \lambda_3^3 = \lambda_1\lambda_5\lambda_8 \quad \lambda_4^3 = \lambda_1\lambda_5^2 \qquad (4\text{-}6)$$

$$\lambda_6^3 = \lambda_5^2\lambda_8 \quad \lambda_7^3 = \lambda_5\lambda_8^2$$

These equations, together with (4-5) determine the solution

$$\lambda_1 = \frac{1+\sqrt{513}}{128} \qquad \lambda_8 = \left(\frac{8\lambda_1 - 1}{6\lambda_1^{2/3}}\right)^3 \approx 0.015$$

Thus, according to the entropy principle about 1.5% of the particle pairs will be detected, in our alleged experiment, as having spin up in both x and y directions on the left and spin down in both y and z directions on the right.

The trouble with such "predictions", of course, is that we do not have a way to perform the experiment. But this is the trouble with realist quantum logic in general, there seems to be no way to isolate the systems which are described by one specific quantum truth function. The existence of quantum truth functions, as depicting actual state of affairs, is thus a mere stipulation.

4.7 Non Realist Quantum Logic

We always have the option to take quantum "truth values" for what they are, limit cases of quantum frequencies, limit cases which sometimes are not obtainable even in theory. Thus $\theta(a) = 1$ does not mean that a is true, rather that almost all systems, from a given source have the property described by a, that is, a vast majority pass the *actual* test for a. The genuine quantum truth function $\theta(a_1) = \theta(a_2) = 1$, $\theta(a_1 \wedge a_2) = 0$ should be interpreted in those terms as well, namely as an idealization of frequencies. It describes a source such that a vast majority of particles (or systems of particles) coming from the source appear to have the property described by a_1, when the test for that property is *actually* performed. A vast majority of the particles coming from the source appear to have the property described by a_2, when the test for that property is performed, and virtually no particle from the source is registered as having property $a_1 \wedge a_2$ when the test for this property is performed. Since every quantum correlation is a convex combination of the quantum truth functions, we can always conceive of every experimental result as a mixture of such extreme cases. This however does not add any new information beyond that which is given by qunatum-theory, namely the frequencies themselves.

In this approach the semantics for quantum logic is not based on the notion of truth but rather on the notion of probability. If we take "probability" as a primitive concept, which is divorced from "truth", then qunatum-logic can be interpreted consistently in terms of extreme probability values: probability 1 (which means "all particles" or "almost all particles") and probability zero (which means "no particle" or "almost no particle"). There is a great difference between "probability 1" and "truth" and we shall explore it in the next chapter.

4.8 Notes and Remarks

The idea that quantum mechanics is best understood in terms of a non-classical logic has been promoted by the philosopher Reichenbach (1944), but the form which is mostly related to my presentation is due to Birkhoff and von Neumann (1943). This approach has been further developed in Jauch (1968), Bub (1974) and from a geometrical perspective in Varadarajan (1962). The version presented here and the results of Section 4.2 were given, without proof in Pitowsky (1986).

The theorem in Section 4.3 is taken from Kochen and Speckler (1967) classic. In this relation see also Pitowsky (1982a), Stairs (1983) and Pitowsky (1985b) for a somewhat different analysis.

The "conspiratorial interpretation" and its implications for locality appear here for the first time. This interpretation is closely related to the "contextual hidden variable" theories, see Gudder (1970) and Shimony (1984). The "operational interpretation" was introduced by Finkelstein (1962), (1968) and by Putnam (1968). For a further criticism of Putnam see Dummet (1978).

Numerous different views exist regarding quantum logic, for a survey see the volume edited by Beltrametti and Van Fraassen (1981).

5. Hidden Variables and Kolmogorovian Models

"The essence of mathamatics lies in its freedom"

George Cantor

5.1 Classical Hidden Variable Theories

In the first chapter we focused on the conceptual gap between the operational formulation of the uncertainty principle and the metaphysical (epistemic or ontic) formulations. The fact that a measurement of the value of one observable A may disturb the value of another B, is not in itself revolutionary, it is compatiable with classical ideas. The metaphysical formulations introduce more radical elements; the epistemic uncertainty principle for example, states that it is impossible to predict from *any* theory what happens to the B value during an A measurement. The family of theories known as (deterministric) hidden variable theories take the opposite view. While the operational uncertainty principle is taken as valid – as are all statistical predictions of quantum mechanics – an attempt is made to assign simultaneous precise value to all observables.

Thus, for example, a hidden variable theory will assign precise spin value (+1/2 or −1/2) to the spin component of an electron in every direction and predict, at least in principle, what precisely happens to the spin component in the x direction during a measurement of the spin in the y direction. In spite of the operational uncertainty relations such theories may have indirect observable consequences which transcend quantum theory.

It is quite often maintained that "deterministic hidden variable theories are impossible" because of this or that mathematical result (such as von Neumann's proof, the violation of Bell inequalities, the Kochen and Specker theorem and so forth). This statement is of course false. The above mentioned results certainly constraint any consistent hidden variable theory, but do not render such theories impossible. Indeed one has to distingish between two questions:

(1) *Are deterministic hidden variable theories logically possible? Or even, are deterministic local hidden variable theories logically possible?* The answer is yes! We shall see various examples of deterministic hidden variable theories which are compatible with quantum statistics. In Sections 5.4, 5.5 I shall develop a *local* deterministic hidden variable theory. The second question is

(2) *Are deterministic hidden variable theories interesting from a physical point of view?* Now this is a more subtle and more serious question because it reflects on our criteria for rational theoretical choice.

In this chapter I shall ignore the second question completely and concentrate only on the logical aspects. This fact must be stressed: The hidden variable theories proposed in this chapter, in particular the Kolmogovonian models of Sections 5.4, 5.5 are not necessarily introduced as physically viable theories. My purpose is, by and large, to point to a logical possibility. In the concluding chapter I shall take up the second question, namely the physical importance of the hidden variable approach.

In classical mechanics the specification of the initial values of a set of physical parameters (say, position and momentum) allow us to predict the state of the system at any furture time. A (deterministic) hidden variable theory is a generic name of a theory which associates with evry physical system a set of possible hidden variables Λ which is the analogue of phase space in classical mechanics. The elements $\lambda \in \Lambda$ are called hidden variable states. We do not assume a–priori that Λ has a structure of a vector space or any other specific mathematical structure. Thus, the elements λ of Λ are themselves sets; each hidden variable state λ may include the position value momentum value angular momentum and spin in various directions and perhaps the value of some yet unknown physical parameters. Given a hidden variable state $\lambda \in \Lambda$ we can calculate, at least in principle, the simultaneous value of all quantum mechanical observables at any future time. In particular the theory can predict what happens to the value of one parameter (say, spin in the x direction) as a result of a measurement of the value of another parameter (spin in the y direction).

Consider a quantum mechanical observable B. The set of all eigenvectors of B, which correspond to a fixed eigenvalue b, is a closed subspace of the Hilbert space associated with the system. In the hidden variable picture the set of all hidden variable states $\lambda \in \Lambda$, in which a measurement of B will give the result b, is a subset of Λ. Hence with every projection E on a closed subspace of a Hilbert space we associate a subset A(E) of Λ, such that λ∈A(E), if and only if a system whose hidden variable state is λ has a the quantum property associated with E.

A hidden variable state λ∈Λ is associated with a precise value of all physical parameters, while the quantum state W (pure or mixture) assign only a probability distribution over possible values. Hence every quantum state W (pure or mixed) is associated with a "mixture" of hidden variable states. For our theory to be consistent we have to be able to calculate from these "mixtures" the probability that a quantum system in the quantum state W has property E, namely tr(WE).

So far the discussion has been most general and abstract. In order to get a detailed theory we have to specify the following:

(1) Which is the set Λ?

(2) What do we mean by a "mixture of hidden variable states"?

(3) What is the nature of the map E → A(E)?

(4) What is the nature of the map W → "mixture of hidden varialbe states"?

(5) How do we calculate the expectation values in the hidden variable theory?

Different answers to these questions yield different theories. The alleged "no hidden variable theorems" in fact point out that particular answers to the above questions are inconsistent. We shall deal with the most important results. *All these results pertain to the case in which by "mixture of hidden variable states" we mean a probability measure on a suitable algebra of subsets of Λ.* So let us introduce:

Definition (5–1): Given a Hilbert space H, a classical hidden variable theory on a set Λ is an association of a subset A(E) ⊂ Λ with every projection E ∈ L(H) and an association of a probability measure μ_w with every state W on H, such that every set A(E) is μ_w–measurable and $\mu_w(A(E)) = tr(WE)$

Classical hidden variable theories surely exist. A trivial example was proposed by Kochen and Specker (1967): Let Λ be the set of all functions $\lambda:L(H) \to \{0,1\}$, where, as usual, L(H) is the lattice of closed orthogonal projections in the Hilbert space H. If E ∈ L(H), let A(E) = $\{\lambda \in \Lambda \mid \lambda(E) = 1\}$. For a state W on H let μ_w be defined on the cylindrical sets in Λ as follows:

$$\mu_w(A(E_1) \cap A(E_2) \cap ... \cap A(E_k)) = \prod_{j=1}^{k} tr(WE_j).$$ Then we can extend μ_w to be a probability measure on Λ defined for all sets in the σ–algebra generated by these cylindrical sets. By definition, $\mu_w(A(E)) = tr(WE)$.

This hidden variable theory makes no sense at all. The reason for its conterintuitiveness is that it ignores completely the relations among observables; all quantum properties are taken to be stochastically independent. Suppose for example that E is the projection on the closed subspace spanned by the energy eignstates for which the energy eigenvalue is ≤ 0.7, then $E^{\perp}$ is the projection on the close subspace spanned by the energy eigenstates corresponding to energy eignvalues > 0.7. Now take a quantum state W such that $tr(WE) \neq 0$ and $tr(WE^{\perp}) \neq 0$. According to the above theory $\mu_w(A(E) \cap A(E^{\perp})) = tr(WE)tr(WE^{\perp}) \neq 0$, hence there is a hidden variable state $\lambda \in A(E) \cap A(E^{\perp})$ which corresponds to a system whose energy value is at once less or equal to 0.7 and greater than 0.7, which is absurd.

So the next natural step to take is to impose certain restrictions on the map $E \to A(E)$.

5.2 Critique of Classical Hidden Variable Theories

We have already seen in Section (3.11) that we cannot possibly perserve all logical relations among observables. To repeat this point we prove:

Lemma (5.1):   There exist no classical hidden variable theory for which the map $E \to A(E)$ satisfies $A(E_1 \wedge E_2) = A(E_1) \cap A(E_2)$ for all $E_1, E_2 \in L(H)$

Proof: Let $n = 2$, $S = \{\{1,2\}\}$, choose projections E_1, E_2 and a state W such that $p_1 = tr(WE_1) = 0.9$, $p_2 = tr(WE_2) = 0.9$ and $p_{12} = tr(W(E_1 \wedge E_2)) = 0$. Then $(p_1, p_2, p_{12}) \notin c(n,S)$, in particular $p_1 + p_2 - p_{12} > 1$. But for any classical hidden variable theory which satisfies $A(E_1 \wedge E_2) = A(E_1) \cap A(E_2)$ we will have $\mu_w(A(E_1)) + \mu_w(A(E_2)) - \mu_w(A(E_1) \cap A(E_2)) = p_1 + p_2 - p_{12} \leq 1$, since μ_w is a probability measure; contradiction.

So the requirement that all logical relations be preserved cannot be satisfied in a classical hidden variable theory. But this, nevertheless, is not a reasonable restriction. Note that the violation of the classical constraints occurs only if we take noncomuting projections E_1, E_2. In this case the simultaneous measurement of E_1, E_2 is anyway impossible and the fact that $A(E_1 \wedge E_2) \neq A(E_1) \cap A(E_2)$ may be explained by reference to measurement disturbances (see Section 3.11). So the next rational step to take is to maintain that the logical relations be preserved only among commuting projections, because in this case all properties involved can be simultaneously determined by a single experiment. But from Kochen and Specker theorem (Section 4.3) we obtain the following:

Theorem (5-2): If H is a Hilbert space of dimension ≥ 3 there exists no classical hidden variable theory for which the map $E \rightarrow A(E)$ satisfies $A(E^{\perp}) = \tilde{A}(E) = \Lambda \backslash A(E)$ for all $E \in L(H)$ and $A(E_1 \wedge E_2) = A(E_1) \cap A(E_2)$ for all commuting pairs E_1, $E_2 \in L(H)$.

Proof: We have seen in Section 4.3 that if x_1, x_2, x_3 are three orthogonal directions in confirguration space and J_{x_1}, J_{x_2}, J_{x3} are the three angular momenta operators in the three dimensional irreducible representation, then *in this particular representation* $J_{x_1}^2$, $J_{x_2}^2$, J_{x3}^2 pairwise communte, and $J_{x_1}^2 + J_{x_2}^2 + J_{x3}^2 = 2I$, (I is the unit operator). Let E_x denote the projection on the one dimensional subspace of the angular momentum space corresponding to the property: "The (square of) the angular momentum in the x direction is zero". Then if x_1, x_2, x_3 is an orthogonal tripple we have: $E_{x_1} \vee E_{x_2} \vee E_{x3} = I$, $E_{x_i} \wedge E_{x_j} = 0$ which entails that $(E_{x_i} \wedge E_{x_j})^{\perp} = I$ for all $1 \leq i < j \leq 3$.

Let x_1, x_2, ..., x_{117} be the directions as in Lemma (4–5) then if $\{a_i, a_j\}$ is an edge in the graph Γ (Fig. (4–2)) we have $(E_{x_i} \wedge E_{x_j})^{\perp} = I$, and if $\{a_i, a_j, a_k\}$ is a triangle in Γ we have $(E_{x_i} \vee E_{x_j} \vee E_{x_k}) = I$. Hence, trivially all the operators in the family $(E_{xi} \wedge E_{xj})^{\perp}$, $\{a_i,a_j\}$ edge in Γ, $E_{x_i} \vee E_{x_j} \vee E_{x_k}$, $\{a_i, a_j, a_k\}$ triangle in Γ pairwise commute and we have:

$$\bigwedge_{\{a_i,a_j\}\ \text{edge in}\ \Gamma} (E_{x_i} \wedge E_{x_j})^{\perp} \wedge \bigwedge_{\{a_i,a_j,a_k\}\ \text{triangle in}\ \Gamma} (E_{x_i} \vee E_{x_j} \vee E_{x_k}) = I \qquad (5\text{–}1)$$

Now suppose that the map $E \to A(E)$ of closed projections to subsets of hidden variable states satisfy $A(E^{\perp}) = \breve{A}(E) = \Lambda \backslash A(E)$, and $A(E_1 \wedge E_2) = A(E_1) \cap A(E_2)$ whenever $[E_1, E_2] = 0$, then if $[E_1, E_2] = 0$ we have by De Morgan rule:

$$A(E_1 \vee E_2) = A[(E_1^{\perp} \wedge E_2^{\perp})^{\perp}] = \breve{A}(E_1^{\perp} \wedge E_2^{\perp}) = A(E_1) \cup A(E_2).$$

Since E_{x_i}, E_{x_j}, E_{x_k} are pairwise orthogonal $A(E_{x_i} \vee E_{x_j} \vee E_{x_k}) = A(E_{x_i}) \cup A(E_{x_j}) \cup A(E_{x_k})$ whenever $\{a_i, a_j, a_k\}$ is a triangle in Γ. Also note that if I is the unit operator then $A(I) = \Lambda$. Since $A(I) = A(E \vee E^{\perp}) = A(E) \cup A(E^{\perp}) = A(E) \cup \breve{A}(E) = \Lambda$. Thus substituting in (5–1) we get

$$\bigcap_{\{a_i,a_j\}\ \text{edge in}\ \Gamma} (\breve{A}(E_{x_i}) \cup \breve{A}(E_{x_j})) \cap \bigcap_{\{a_i,a_j,a_k\}\ \text{triangle in}\ \Gamma} (A(E_{x_i}) \cup A(E_{x_j}) \cup A(E_{x_k})) = \Lambda$$

But this is a contradiction. To see this take $\lambda \in \Lambda$ and define a classical truth function $t : \{a_1, a_2, \ldots a_{117}\} \rightarrow \{0,1\}$ by $t(a_i) = 1$ iff $\lambda \in A(E_{x_i})$ then:

$$t\left[\bigwedge_{\{a_i, a_j\} \text{ edge in } \Gamma} [\sim(a_i \wedge a_j)] \quad \wedge \quad \bigwedge_{\{a_i, a_j, a_k\} \text{ triangle in } \Gamma} (a_i \vee a_j \vee a_k)\right] = 1$$

while the proposition above is a classical logical contradiction!

Theorem (5–2) has a rather limited scope because of the assumption $A(E^{\perp}) = \tilde{A}(E)$. This assumption is nothing but the identification of $E^{\perp}$ with the property "not E" and we have already seen in Section 3.3 that such a move is problematic. While the operation $E_1 \wedge E_2$ does correspond to the intersection of two subspaces, and therefore can legitmately be taken as the classical logical "and", the orthocomplemetation operation has no similar set theoretic interpretation. Kochen and Specker result strengthens our suspicion in that regard.

The most important restriction on any classical hidden variable theory is associated with the violation of the classical constraints on correlations in the E.P.R. experiment. I have argued this point at length in Section 3.11 and here I shall only briefly repeat the point.

Theorem (5–3): There exists no classical local hidden variable theory.

Proof: Take $n = 4$, $S = \{\{1,3\}, \{1,4\}, \{2,3\}, \{2,4\}\}$. Consider the four projections E_1, E_2, E_3, E_4 in formula (3–12) and the singlet state W_s. We have

$p_i = tr(W_s E_i) = 1/2$ $1 \leq i \leq 4$ and $p_{ij} = tr[W_s(E_i \wedge E_j)] = 3/8$ for $\{i,j\} = \{1,3\}$,

$\{1,4\}$, $\{3,4\}$ and $p_{23} = tr[W_s(E_2 \wedge E_3)] = 0$. Now suppose that we had a map

$E \to A(E)$ and $W_s \to \mu$ of a classical hidden variable theory, then since

$p = (p_1, p_2, p_3, p_4, p_{13}, p_{14}, p_{23}, p_{24})$ is not an element of $c(4,S)$, we must have

$A(E_i \wedge E_j) \neq A(E_i) \cap A(E_j)$ for some $\{i,j\} \in S$. Suppose without loss of generality

that $i = 1$, $j = 3$, then choose $\lambda \in \Lambda$ such that $\lambda \in A(E_1 \wedge E_3)$, $\lambda \notin A(E_1) \cap A(E_3)$

(other cases are dealt with in a similar manner.) If we had a source of singlet

state pairs, *all* in the hidden variable state λ, we would be able to transmit a

superluminal message with high probability, as I have already argued before (in

Sections 3.11 and 4.4). Hence the hidden variable states are non local.

It must be stressed again that classical hidden variables theories do not

necessarily maintain that one can achieve superluminal signaling in practice. We

may not be able to control the hidden variable states. The important point is that

the states λ themselves are non–covariant and thus any realistic assumption

regarding their existence contradicts the theory of relativity.

5.3 Non-Classical Hidden Variables – The Geometric Analogy

What makes the classical hidden variable theories "classical" is the

identification of "mixtures of hidden variable states" as probability measures on the

set of hidden variables Λ (definition (5–1)). All difficulties associated with the

classical hidden variable approach stem from this fact, most notably the failure of

locality (theorem (5–3)). The logical step to take in this circumstance is to extend the notion of "probabilty measure" to accomdate the situation. One way to go about it is to use (cheap) tricks such as negative "probability", or even complex "probability" values. Formally we may be able to "solve" our problem, but then the term "probability" loses completely its meaning. In particular our basic "balls in an urn" model, which underlies the axioms of probability, no longer makes sense. It is absurd to talk about an urn containing -17 red balls or $3e^{i\pi/12}$ wooden balls.

Our aim is therefore twofold:

1. To extend the notion of probability while retaining the basic intuition of the "balls in an urn" model, that is, the idea that probability distributions reflect proportions of different properties, and is consequently reflected in relative frequencies.

2. To construct a local hidden variable theory based on this extention.

The situation is somewhat analoguous to the case of non-Euclidean geometrics. After centuries of futile attemtps to derive Euclid's fifth postulate (the axiom of parallels) from the other four postulates, Lobachewsky proposed to take more seriously the idea that this might be impossible. He assumed the validity of one of the possible negations of the axioms of parallels (namely, that there are infinitely many lines parallel to a given line through a given point in a plane), and obtained the so-called "hyperbolic geometry".

No contradiction seemed to arise from Lobachewsky's assumption, the new hyperbolic geometry appeared to be perfectly consistent, though Lobachewsky was not able to demonstrate that explicitely. But still the game was counterintuitive, no one could "see" or imagine what hyperbolic lines "really" are, and how it is

149

possible that the sum of the angles in a triangle is less then 180°. In other words, no models of hyperbolic geometry were known to exist. The first to construct such a model explicitly was Kline almost 50 years after Lobachewsky. His model was followed by other models, most notably the one due to Poincaré. A few features of these models should be mentioned in the present context:

a. The models are called "Euclidean models of hyperbolic geometry" becuase hyperbolic straight lines are either segments of Euclidean lines (Kline) or segments of Euclidan curves (Poincaré).

b. Hence there is no unique concept of "hyperbolic line" but rather many differnt realizations.

c. The hyperbolic "distance" is different from the usual Euclidean metric, but is nevertheless defined in terms of the latter.

Quantum probability is presently in a state analogous to the state of hyperbolic geometry after Lobachewsky but before Kline. We have a theory – quantum mechanics – which allows us to calculate probabilities of a large variety of microphysical events, but we still cannot "see" or imagine how the predicted and observed frequencies could arise from a probability distribution in the usual sense. What we lack, in other words, is a "Kolmogorovian model of quantum probability" which is classical in the sense that quantum events are depicted as subsets of a given set, so that the family of all events forms a Boolean algebra, and the logical relations among quantum observables are maintained. And is not classical in the sense that the "properties", that is subsets of Λ, are of an unusual kind (in analogy with the hyperbolic lines, which are not Euclidean straight lines); and the "probability measure" is not a usual distribution, but is nevertheless defined in terms of a classical probability measure (in analogy with the hyperbolic metric).

Surely such a "Kolmogorovian model of quantum probability" cannot be constructed on a finite set or even a countable set. It takes some "paradoxical" features of the continuum to conceive of a model which will recover the values of quantum frequencies in a local way.

5.4 Kolmogorovian Models of Quantum Statistics: Mathematical Introduction

The Lebesgue measure of a subset of the straight line R is an extension of the concept of "length" to that subset. The question of whether any arbitrary subset of R can be assigned "length" in a consistent manner has no defintie answer, that is the answer depends on the validity of the axiom of choice (more on that in Section 5.6). If however, we accept the axiom of choice as valid, we can prove that there are many subsets of R to which "length" cannot be assigned consistently. We shall encounter examples shortly. Similarly the axiom of choice entails that there are amny subsets of the plane $R^{(2)}$ or space $R^{(3)}$ to which no "area" and "volume" can be assigned consistently. The latter case, $R^{(3)}$ provides perhaps the most dramatic example, the so-called Banach–Tarski paradox:

"It is possible to decompose a solid ball into eight subsets which can be so rearranged (by rotations and translations) as to obtain two solid balls of the same size as the original one."

In spite of the name, this is really not a paradox. The subsets into which the ball is decomposed are nonmeasurable subsets in $R^{(3)}$, that is we cannot assign them numbers which indicate their volume without violating the rotational and translational invariance of "volume."

Our "events" will be peculiar subsets of the unit interval [0,1], which have similar "paradoxical" character. Let m be the Lebesgue measure on the interval [0,1] and let Σ_0 be the σ-algebra of Lebesgue measurable subsets of [0,1], then ([0,1], Σ_0, m) is a classical probability space. For two numbers r_1, $r_2 \in$ [0,1] let $r_1 + r_2$ denote the addition *modulo 1* (that is $r_1 + r_2$ is just the usual sum in case this sum is less than 1, and it is the usual sum of r_1 and r_2 minus 1 in case the sum exceeds or equal 1). With this operation [0,1] is turned into an abelian group (the circle group) and the Lebesgue measure is +-invariant, that is

$$m(A + r) = m\{x + r| \ x \in A\} = m(A) \text{ for all } r \in [0,1] \text{ and } A \in \Sigma_0.$$

Definition (5-2): Let $A \subset [0,1]$ be an arbitrary set, which is not necessarily Lebesgue measurable, the outer measure of A is:

$$\overline{m}(A) = \inf \{m(G)| \ G \supseteq A, \ G \text{ open}\} \qquad (5-2)$$

A subset $G \subseteq [0,1]$ is open iff it is a countable union of disjoint open intervals. The concept "outer measure of A" can therfore be explained in the following way: We have a large collection of rulers of practically every conceivable size. Given a set $A \subseteq [0,1]$ we cover it by rulers from our collection until no point of A can be seen. The sum total of the lengths of the covering rulers is a first approximation to the "size" of A. Next we improve the cover by taking smaller rulers which still cover A, and again compute the sum total of lengths. We continue in this refinement. The greatest lower bound on the lengths obtained in this way is the outer measure of A. Since for every open set $G \subseteq [0,1]$ the set

$G + r = \{x + r \mid x \in G\}$ is also open (remeber $x + r$ denotes addition modulo 1), the outer measure m is also $+$-invariant, $\bar{m}(A + r) = \bar{m}(A)$. For $A \subseteq [0,1]$, let $\tilde{A}$ denote its complement in $[0,1]$, that is $\tilde{A} = [0,1]\backslash A$, then the *inner measure* of A, $\underline{m}(A)$ is:

$$\underline{m}(A) = 1 - \bar{m}(\tilde{A}) \qquad\qquad (5\text{--}3)$$

Since a set $F \subseteq [0,1]$ is closed iff $\tilde{F}$ is open we have

$$\underline{m}(A) = \sup \{m(F) \mid F \subseteq A, F \text{ closed}\} \qquad\qquad (5\text{--}4)$$

If A is Lebesgue measurable then $\bar{m}(A) = m(A) = \underline{m}(A)$. The following theorem establishes the existence of a vast number of nonmeasurable sets and is in a sense even more striking then the Banach–Tarski paradox.

Theorem (5–4)   There exists a decomposition of the interval [0,1] into a contiuum (that is, $2^{\aleph_0}$) of pairwise disjoint sets each having outer measure 1 (and inner measure zero).

Proof: Let $\mathscr{F}$ be the family of all compact subsets of $[0,1]$ whose cordinality is $2^{\aleph_0}$. Since the complement of each $F \in \mathscr{F}$ is an open set, and since every open subset of $[0,1]$ is a countable union of disjoint open intervals, it follows that every $F \in \mathscr{F}$ is determined by a countable set (the edges of the intervals in $\tilde{F}$). Hence $|\mathscr{F}| = 2^{\aleph_0}$. By the axiom of choice, or rather its equivalent "principle of well ordering", we can well order $\mathscr{F}$, that is $\mathscr{F} = \{F_\alpha \mid \alpha < \Omega\}$ where Ω is the least ordinal number whose cardinality is $2^{\aleph_0}$. Let F be a compact subset of $[0,1]$ such that $m(F) > 0$, then F contains a non–empty perfect subset (in fact F is a union of a perfect set and a countable set). Therefore F contains $2^{\aleph_0}$ compact subsets whose cardinality is $2^{\aleph_0}$. It follows that for every ordinal $\alpha < \Omega$ there exists $\alpha \leq \beta < \Omega$ such that $F_\beta \subseteq F$.

Let $\{r_\alpha \mid \alpha < \Omega\}$ be a well ordering of all the numbers in $[0,1)$. For a fixed $\alpha < \Omega$ and $x \in [0,1)$ let $c_\alpha(x)$ denote all expressions of the form $\pm r_{\beta_1} \pm r_{\beta_2} \pm \ldots \pm r_{\beta_n} \pm x$, where $1 \leq \beta_k \leq \alpha$, $k = 1, 2, \ldots, n$ and $n = 1, 2 \ldots$, the addition is, as usual, taken to be modulo 1. Then $|c_\alpha(x)| = \max(\aleph_0, |\alpha|) < 2^{\aleph_0}$.

We shall now prove: "There exists a double transfinite sequence $\{x_\beta^\alpha \mid 1 \leq \beta \leq \alpha < \Omega\}$ of numbers in the interval $[0,1)$ such that

(a) $x_\beta^\alpha \in F_\alpha$ $1 \leq \beta < \alpha < \Omega$

(b) The sets $c_\alpha(x_\beta^\alpha)$ for $1 \leq \beta \leq \alpha < \Omega$ are pairwise disjoint.

To see that order the ordinal pairs by lexiographic order $(\gamma,\delta) < (\alpha,\beta)$ if $\gamma < \alpha$ or $\gamma = \alpha$ and $\delta < \beta$. With this order the set $\{(\alpha, \beta)|\ 1 \leq \beta \leq \alpha < \Omega\}$ is well ordered. Now define the set $\{(x_\beta^\alpha)\ |\ 1 \leq \beta \leq \alpha < \Omega\}$ by induction: Let x_1 be an arbitrary element of F_1. Suppose that $1 \leq \beta \leq \alpha < \Omega$ and suppose that x_δ^γ where chosen for every pair $(\gamma,\delta) < (\alpha,\beta)$ for which $1 \leq \delta \leq \gamma$. Now put

$$D(\alpha,\ \beta) = \bigcup_{(\gamma,\delta) < (\alpha,\beta)} c_\alpha(x_\delta^\gamma)$$

Then $|D(\alpha,\beta)| \leq |\alpha|^2 \max(\aleph_0, |\alpha|) < 2^{\aleph_0}$. Hence $\breve{D}(\alpha,\ \beta) \cap F_\alpha \neq \phi$, where $\breve{D}(\alpha,\beta) = [0,1)\backslash D(\alpha,\beta)$. Chose an arbitrary element $x_\beta^\alpha \in \breve{D}(\alpha,\beta) \cap F_\alpha$. Then $c_\alpha(x_\beta^\alpha)$ is disjoint from every $c_\gamma(x_\delta^\gamma)$ for $(\gamma,\delta) < (\alpha,\beta)$, otherwise we would have, by definition

$$\pm\ r_{\beta_1} \pm r_{\beta_2} \pm \ldots \pm r_{\beta_n} \pm x_\beta^\alpha = \pm r_{\delta_1} \pm r_{\delta_2} \pm \ldots \pm r_{\delta_m} \pm x_\delta^\gamma$$

where $1 \leq \beta_k \leq \alpha$, $1 \leq k \leq n$, $1 \leq \delta_j \leq \gamma$, $1 \leq j \leq m$. Hence

$$x_\beta^\alpha = \pm\ r_{\delta_1} \pm r_{\delta_2} \pm \ldots \pm r_{\delta_m} \pm r_{\beta_1} \pm r_{\beta_2} \pm \ldots \pm r_{\beta_m} \pm x_\delta^\gamma \in c_\alpha(x_\delta^\gamma) \subseteq D(\alpha,\ \beta)$$

contrary to the choice of x_β^α.

Let $1 \leq \nu < \Omega$ be an ordinal and define:

$$X_\nu = \bigcup_{\nu \leq \alpha < \Omega} c_\alpha(x_\nu^\alpha)$$

Then $|X_\nu| = 2^{\aleph_0}$, also $X_\nu \cap X_\sigma = \phi$ for $\nu \neq \sigma$ by construction. Moreover

$\overline{m}(X_\nu) = 1$; for if $\overline{m}(X_\nu) < 1$, then there exists an open set G such that $G \supseteq X_\nu$

and $m(G) < 1$. Let $F = \hat{G} = [0,1) \backslash G$ then $m(F) > 0$. Hence by the

remark at the beginning of the proof there is $\nu \leq \alpha < \Omega$ such that $F_\alpha \subseteq F \subseteq \hat{X}_\nu$.

But this is a contradiction since $x_\nu^\alpha \in F_\alpha$ by construction and also

$x_\nu^\alpha \in c_\alpha(x_\nu^\alpha) \subset X_\nu$. Therefore $\{X_\nu \mid 1 \leq \nu < \Omega\}$ is a family of $2^{\aleph_0}$ pairwise disjoint

subsets of $[0,1]$, each having outer measure 1.

We shall introduce now the notion of "mixture" appropriate for our purposes.

Definition (5-3): Let Σ be a σ-algebra of subsets of $[0,1]$. A function

$\rho: \Sigma \to [0,1]$ is called a mixture on Σ if it satifies

(a) ρ is monotone $\rho(A) \leq \rho(B)$ whenever $A \subset B$, $A,B \in \Sigma$,

(b) $\underline{m}(A) \leq \rho(A) \leq \overline{m}(A)$ for all $A \in \Sigma$.

Given a mixture ρ on Σ, two sets A, B $\in \Sigma$ are ρ-comeasurable if

$\rho(A) + \rho(B) = \rho(A \cup B) + \rho(A \cap B)$.

Note that if $A \in \Sigma$ is Lebesgue measurable and ρ a mixture on Σ then

$\rho(A) = m(A)$. In particular $\rho([0,1]) = 1$, $\rho(\phi) = 0$ for every mixture ρ.

The motivation behind this definition has to do with an extension of the law of large numbers. Let us first examine what this law asserts for measurable sets. Let $A \subseteq [0,1]$ and suppose that we sample points from the interval $[0,1]$ at random, what is the frequency of A-points in the sample? If A is Lebesgue measurable the answer is clear: "With probability 1 the frequency of A-points approaches $m(A)$ as the sample grows larger". This is the content of the law of large numbers which we shall formulate and prove now. Let $n \geq 1$ be a natural number and let $[0,1]^n = [0,1] \times ... \times [0,1]$ (n-copies) be the space of all n-tupples of elements from $[0,1]$. Let M_n be the Lebesgue measure on $[0,1]^n$, that is for every set of the form $A_1 \times A_2 \times ... \times A_n \subseteq [0,1]^n$, where the A_j's are Lebesgue measurable, we put

$$M_n(A_1 \times ... \times A_n) = \prod_{j=1}^{n} m(A_j),$$ and then we extend M_n to be defined on the σ-algebra generated by these cylindrical sets.

For a fixed Lebesgue measurable set $A \subseteq [0,1]$ let χ_A be the indicator function of A, $\chi_A(x) = 1$ if $x \in A$, and $\chi_A(x) = 0$ for $x \notin A$. For $(x_1, ..., x_n) \in [0,1]^n$ the number $n^{-1} \sum_{i=1}^{n} \chi_A(x_i)$ is the frequency of A-points in the sample $(x_1, ..., x_n)$. We shall prove

<u>Lemma (5-5)</u> (The Classical Law of Large Numbers): Let $A \subset [0,1]$ be Lebesgue measurable then the set:

$$F(A,n,\varepsilon) = \{(x_1, x_2 ... x_n) \in [0,1]^n; |\ n^{-1} \sum_{i=1}^{n} \chi_A(x_i) - m(A)| \geq \varepsilon\} \quad (5-5)$$

is M_n-measurable and $M_n(F(A,n,\varepsilon)) \leq \dfrac{10}{n^2 \varepsilon^4}$

Proof: For the set $A \subseteq [0,1]$ put $A^1 = A$, $A^0 = \widehat{A} = [0,1]\backslash A$. Let

$B_k = \{(x_1, ..., x_n) \in [0,1]^n; \sum_{i=1}^{n} \chi_A(x_i) = k\}$ for $k = 0, 1, ..., n$. Then

$B_k = \cup(A^{\varepsilon_1} \times A^{\varepsilon_2} \times ... A^{\varepsilon_n})$ where the union is taken over all $\binom{n}{k}$ sequences

$(\varepsilon_1, ..., \varepsilon_n) \in \{0,1\}^n$ such that $\sum \varepsilon_i = k$. Hence $B_k \subseteq [0,1]^n$ is M_n-measurable and

$M_n(B_k) = \binom{n}{k}[m(A)]^k[1-m(A)]^{n-k}$. Put $p = m(A)$, $q = 1-m(A)$

then $F(A,n,\varepsilon) = \cup B_k$ where the union is taken over all k such that

$|\frac{k}{n} - p| \geq \varepsilon$. Hence $F(A,n,\varepsilon)$ is M_n-measurable and

$$M_n(F(A,n,\varepsilon)) = \sum_{|k-np|\geq n\varepsilon} M_n(B_k) \leq \sum_{|k-np|\geq n\varepsilon} (\frac{k-np}{n\varepsilon})^4 M_n(B_k)$$

$$\leq \sum_{k=0}^{n}(\frac{k-np}{n\varepsilon})^4 M_n(B_k) = \frac{1}{n^4\varepsilon^4} \sum_{k=0}^{n} (k- np)^4 \binom{n}{k}p^k q^{n-k} =$$

$$= \frac{npq(3npq-6pq+1)}{n^4\varepsilon^4} \leq \frac{10}{n^2\varepsilon^4} \text{ for } n \geq 2$$

From Lemma (5–5) it follows that $M_n(F(A,n,\varepsilon)) \to 0$ for every $\varepsilon > 0$,

$M_n(\widehat{F}(A,n,\varepsilon)) \to 1$ where $\widehat{F}(A,n,\varepsilon)$ is the complement of $F(A,n,\varepsilon)$ in $[0,1]^n$. This

means that for every fixed error $\varepsilon > 0$, the probability that the frequency of

A–points in a sample $(x_1 ... x_n)$, will be within distance ε from the expectation

$m(A)$, approaches 1 as the sample grows.

Our purpose is to extend these results to nonmeasurable sets. First note that

the concepts of the "outer measure" and "inner measure" extend immediately to

$[0,1]^n$ when we define $\overline{M}_n(A) = \inf\{M_n(G) \mid A \subseteq G, G \text{ open}\}$ and

$\underline{M}_n(A) = 1 - \overline{M}_n(\widehat{A})$ for $\widehat{A} = [0,1]^n\backslash A$. We shall prove now

Lemma (5-6):

(a) Let $A \subseteq [0,1]$ then there are Lebesgue mesurable sets F,G such that $F \subset A \subset G$, $m(F) = \underline{m}(A)$, $m(G) = \overline{m}(A)$;

(b) Let $A,B \subset [0,1]$ and let G,H be Lebesgue measurable with $A \subset G$, $\overline{m}(A) = m(G)$, $B \subset H$, $\overline{m}(B) = m(H)$ then $\overline{m}(A \cup B) = m(G \cup H)$

Proof: (a) By definition of outer measure for every natural number n there exists an open set G_n such that $A \subseteq G_n$ and $m(G_n) - \dfrac{1}{n} \leq \overline{m}(A) \leq m(G_n)$, put $G = \overset{\infty}{\underset{n=1}{\cap}} G_n$, then G is measurable (note that G is not necessarily open, it is a so-called G_δ set) and surely $m(G) = \overline{m}(A)$. Now take H to be a set such that $\widetilde{A} \subseteq H$, $\overline{m}(\widetilde{A}) = m(H)$ and put $F = \widetilde{H}$, then $F \subseteq A$, $m(F) = 1 - m(\widetilde{H}) = 1 - \overline{m}(\widetilde{A}) = \underline{m}(A)$.

(b) Let L be a Lebesgue measurable set such that $A \cup B \subseteq L$, $\overline{m}(A \cup B) = m(L)$. Then $A \subseteq G \cap L$ and hence $m(G) = \overline{m}(A) \leq m(G \cap L) \leq m(G)$ and therefore $m(G) = m(G \cap L)$, by the same argument $m(H) = m(H \cap L)$ hence

$$m(G \cup H) = m(G) + m(H) - m(G \cap H) = m(G \cap L) + m(H \cap L) -$$

$$m(G \cap H) \leq m(G \cap L) + m(H \cap L) - m(G \cap H \cap L) = m[(G \cup H) \cap L]$$

$$\leq m(L) = \overline{m}(A \cup B)$$

But since $A \cup B \subseteq G \cup H$ we have $\overline{m}(A \cup B) \leq m(G \cup H)$ hence

$\overline{m}(A \cup B) = m(G \cup H)$.

Note that Lemma (5-6) remains valid when we replace the interval [0,1] and the Lebesque measure m, by the hypercube $[0,1]^n$ and the Lebesque measure M_n on it. We shall now prove the law of large numbers for non-measurable sets:

Theorem (5-7): Let $A \subset [0,1]$ be an arbitrary set

(a) If r is a real number, $\underline{m}(A) \leq r \leq \overline{m}(A)$, then for every $\varepsilon > 0$

$$\overline{M}_n\{(x_1, ..., x_n) \in [0,1]^n; |n^{-1} \sum_{i=1}^{n} \chi_A(x_i) - r| < \varepsilon\} \to 1;$$
$$n \to \infty$$

(b) If r is a real number $0 \leq r < \underline{m}(A)$ or $\overline{m}(A) < r \leq 1$ then

$$\overline{M}_n\{(x_1, ..., x_n) \in [0,1]^n; |n^{-1} \sum_{i=1}^{n} \chi_A(x_i) - r| \leq \varepsilon\} \to 0,$$
$$n \to \infty$$

for all sufficiently small $\varepsilon > 0$

Proof: (a) Denote $K(A,n,r,\varepsilon) = \{(x_1, ..., x_n) \in [0,1]^n; |n^{-1} \sum_{i=1}^{n} \chi_A(x_i) - r| \leq \varepsilon\}$

then we have $K(A,n,r,\varepsilon) = \cup B_k$ where $B_k = \cup(A^{\varepsilon_1} \times A^{\varepsilon_2} \times ... \times A^{\varepsilon_n})$, the

union taken over all $(\varepsilon_1, \varepsilon_2, ..., \varepsilon_n) \in \{0,1\}^n$ for which $\Sigma \varepsilon_j = k$ (as before,

$A^1 = A$, $A^0 = \overline{A}$). Now, let F, G be two measurable subsets of [0,1] such that

$F \subseteq A \subseteq G$ $\overline{m}(A) = m(G)$ and $\underline{m}(A) = m(F)$, now put $L^1 = G$ and $L^0 = \overline{F}$ then

$A^{\varepsilon_1} \times A^{\varepsilon_2} \times ... \times A^{\varepsilon_n} \subseteq L^{\varepsilon_1} \times L^{\varepsilon_2} \times ... \times L^{\varepsilon_n}$. The set on the right hand side is

---

M_n-measurable and it is clear that $\overline{M}_n(A^{\varepsilon_1} \times A^{\varepsilon_2} \times ... \times A^{\varepsilon_n}) =$

$M_n(L^{\varepsilon_1} \times L^{\varepsilon_2} \times ... \times L^{\varepsilon_n})$ since no M_n-measurable set of smaller measure can

contain $A^{\varepsilon_1} \times A^{\varepsilon_2} \times ... \times A^{\varepsilon_n}$

By Lemma (5-6) part (b) applied to the sets $A^{\varepsilon_1} \times A^{\varepsilon_2} \times ... \times A^{\varepsilon_n}$ we have:

$$\overline{M}_n(K(A,n,r,\varepsilon)) = M_n[\bigcup(L^{\varepsilon_1} \times L^{\varepsilon_2} \times ... \times L^{\varepsilon_n})]$$

where, as before, the union is taken over all $(\varepsilon_1, ..., \varepsilon_n) \in \{0,1\}^n$ such that

$\sum\limits_{j=1}^{n} \varepsilon_j = k$ for all k such that $|k - nr| \leq n\varepsilon$.

Given $\underline{m}(A) \leq r \leq \overline{m}(A)$ let $F \subseteq R \subseteq G$ be a measurable set with

$m(R) = r$ then $R \subseteq G$ and $\overset{\frown}{R} \subseteq \overset{\frown}{F}$ hence

$L^{\varepsilon_1} \times L^{\varepsilon_2} \times ... \times L^{\varepsilon_n} \supseteq R^{\varepsilon_1} \times R^{\varepsilon_2} \times ... \times R^{\varepsilon_n}$ where $R^1 = R$, $R^0 = \overset{\frown}{R}$, therefore

$\overline{M}_n(K(A,n,r,\varepsilon)) \geq M_n[\cup(R^{\varepsilon_1} \times R^{\varepsilon_2} \times ... \times R^{\varepsilon_n})]$ with the union taken as before.

Hence

$$\overline{M}_n(K(A,n,r,\varepsilon)) \geq \sum\limits_{|k-nr|<\varepsilon} \binom{n}{k}[m(R)]^k[\overset{\frown}{m}(R)]^{n-k} =$$

$$= \sum\limits_{|k-nr|<\varepsilon} \binom{n}{k}r^k(1-r)^{n-k} \geq 1 - \frac{10}{n^2\varepsilon^4} \underset{n\to\infty}{\to} 1$$

By the law of large number, Lemma (5-5).

(b) Suppose that $\overline{m}(A) < r \leq 1$, we shall show that for all

$0 < \varepsilon < \frac{1}{2}[r - \overline{m}(A)]$ we have $\overline{M}_n[K(A,n,r,\varepsilon)] \underset{n\to\infty}{\to} 0$. Suppose by negation that

$\lim\limits_{n\to\infty} \sup \overline{M}_n[K(A,n,r,\varepsilon)] = \Delta > 0$. Let G be measurable, $A \subseteq G$, $\overline{m}(A) = m(G)$, then

by the classical law of large numbers

$$M_n\{(x_1, ..., x_n) \in [0,1]^n \; ; \; |n^{-1} \sum_{i=1}^{n} \chi_G(x_i) - m(G)| < \varepsilon\} \underset{n\to\infty}{\to} 1$$

Hence there exists an index n such that the set

$$\{(x_1, ..., x_n) \in [0,1]^n \; ; \; |n^{-1} \sum_{j=1}^{n} \chi_G(x_i) - m(G)| < \varepsilon\} \cap K(A,n,r,\varepsilon)$$

is non empty. Let $(x_1, ..., x_n)$ be in this set, then since $A \subseteq G$ we have $\chi_A(x) \leq \chi_G(x)$ for all $x \in [0,1]$ and thus:

$$r - \varepsilon \leq \frac{1}{n} \sum_{i=1}^{n} \chi_A(x_i) \leq \frac{1}{n} \sum_{i=1}^{n} \chi_G(x_i) \leq m(G) + \varepsilon = \overline{m}(A) + \varepsilon$$

hence $r - \overline{m}(A) \leq 2\varepsilon$ contrary to the choice of ε. Therefore

$\overline{M}_n[K(A,n,r,\varepsilon)] \underset{n\to\infty}{\to} 0$ for $\overline{m}(A) < r \leq 1$ and all $0 < \varepsilon < \frac{1}{2}[r - \overline{m}(A)]$. If

$0 \leq r < \underline{m}(A)$ we can prove the claim by interchanging A with $\hat{A}$.

From the theorem it follows directly that if A is nonmeasurable then

$$\underline{M}_n\{(x_1, ..., x_n) \in [0,1]^n; |n^{-1} \sum_{i=1}^{n} \chi_A(x_i) - r| < \varepsilon\} \underset{n\to\infty}{\to} 0 \text{ for all } \varepsilon > 0, \text{ and all}$$

$0 \leq r \leq 1$. For if $\underline{m}(A) \leq r \leq \overline{m}(A)$, then we obtain from part (a) of the theorem that both $K(A,n,r,\varepsilon)$ and its complement have outer measure tending to 1. For all other values of r the claim follows from part (b) of the theorem.

5.5 Kolmogorovian Models of Quantum Statistics

We shall construct a local hidden variable theory (albeit, not a classical one) along the following principles:

Principle I: With every physical system we associate, at every given moment a hidden variable state λ, which is just a number between zero and one. The set of all hidden variable states Λ is thus the interval [0,1]. The value of the hidden variable determines, among other things, all the values of the quantum mechanical observables associated with the system, and all their future values. With every physical property we associate a subset A ⊆ [0,1] such that the particle has the said poperty if and only if its hidden variable state λ is an element of A. The set of all properties forms a σ-algebra Σ.

Principle II: The statistical state of an ansamble of particles, whose precise hidden variable states are unknown, is given by a mixture ρ on Σ (definition (5-3)).

Principle III: Given an ansamble of particles in the statistical state ρ, the following rule obtains: If A,B are ρ-comeasurable properties, that is
$$\rho(A \cup B) + \rho(A \cap B) = \rho(A) + \rho(B) \text{ (defintion (5-3)), then the existence of both}$$
properties A,B can be simultaneously verified by a single experiment on a single sample. If, however, A,B are not ρ-comeasurable we cannot verify the existence of

the properties A,B by a single experiment on a single sample, rather we can measure for the property A on one smaple and for B on a different sample.

Principle IV: In a random sample of particles, whose statistical state is given by the mixture ρ, the frequencey of particles having property A approaches $\rho(A)$ as the sample grows.

It is this last principle which is the most subtle. Recall the law of large numbers for non measurable sets (theorem (5-7)). It states the following:

"For all $\varepsilon > 0$, and $\underline{m}(A) \leq r \leq \bar{m}(A)$, the outer measure of the set of points $(x_1, ..., x_n) \in [0,1]^n$ for which the frequency $n^{-1} \sum_{i=1}^{n} \chi_A(x_i)$ falls within distance ε from r, approaches 1 as n tends to infinity. If $\bar{m}(A) < r \leq 1$ or $0 \leq r < \underline{m}(A)$ the outer measure tends to zero as the sample grows".

Suppose that we sample points from the interval [0,1] at random and ask: What is the limit of frequency of A–points in the sample? By the above law of large numbers any number r, $\underline{m}(A) \leq r \leq \bar{m}(A)$ is an equally likely candidate, and in fact, the frequencies may have no limit if A is nonmeasurable. (The frequencies can oscilate between two values r_1, r_2 with $\underline{m}(A) \leq r_1 < r_2 \leq \bar{m}(A)$, say). But given a finite sample *some* frequency will be observed. Princple IV states that this frequency is approaching $\rho(A)$ as the sample grows. *This is not a mathematical theorem, it is merely a consitent claim.* We simply maintain that from all possible observed frequencies the one which is *actually* realized tend to $\rho(A)$. Since $\underline{m}(A) \leq \rho(A) \leq \bar{m}(A)$ this assumption is mathematically consistent.

Let us examine how these principles operate together in a hypothetical case. Suppose that we have a source which emits particles and we consider two properties, call them "red" and "small". The property "red" is associated with a set $A \subseteq [0,1]$, that is a particle is red iff its hidden variable state λ is an element of A. The property "small" is associated with a subset $B \subseteq [0,1]$. Assume that all the particles in the source are in the statistical state given by the mixture ρ. If A,B, are ρ-comeasurable then we check directly on a single sample how many particles in the sample have both property A and property B.

Suppose that A,B, are not ρ-comeasurable. In this case we determine the frequency of A-particles on one sample, the frequency of B-particles on another sample, the frequency of $A \cap B$-particles on a third sample and so forth. Suppose, for example, that $\rho(A) = 0.9$, $\rho(B) = 0.95$, $\rho(A \cap B) = 0$ (to see how this is possible take two disjoint sets A,B each having outer measure 1, as in theorem (5-4), then the inner measure of both A and B is zero and thus the values of ρ are consistent). Hence, by principle IV, the observed frequency of A-particles in one sample has value close to 0.9, the observed frequency of B-particles in a second sample is 0.95 and the observed frequency of $A \cap B$-particles in a third sample is 0.

This seems counterintuitive and indeed it has been a source of misunderstanding and objection (for example by Mermin (1982) and Mackdonald (1982) in relation to another similar hidden variable theory, Pitowsky (1982b, 1983). The objection goes as follows: We measure "red" on one sample and discover that 90% of the sampled particles have that property. We would have liked to conclude: *"About 90% of the particles in the source are red"*. But this conclusion is unwarranted, even from the point of view of classical probability. The observed result may have been a statistical fluke. The only thing which we can safely say is: *"With high probability, about 90% of the particles in the source are red"*, this is just the law of large numbers. Now we take another sample, measure for the property "small", and find that 95% of the particles in the sample are small and conclude that with high probability this is also the percentage of small particles in the source. We measure "red and small" on a third sample, find no particle with that property in the sample. Shouldn't we conclude that at least one of the samples was "wildly non-random"?

Well, there are two ways to play the game, the right way and the wrong way. The wrong way is to use the term "probability" inconsistantly and this is the source of the objection. The expectation values 90% for red, 95% for small, 0% for red and small are dervied from a non classical concept of probability call it "probability I". The claim that the statistical results indicate, that at least one of the samples is wildly non–random, is a result of the use of the classical concept of probability, that is "probability II". What Mermin and Mackdonald indicate is that the statement: "with high *probability II*, 90% of the particles in the source are red, 95% are small, and 0% are red and small" is contradictory. This is true enough but irrelvant.

The relevant statment is the one in which "propability I" appears instead of "probability II". This forumulation represents a consistent use of the term probability both for expectation values and for frequencies. The law of large numbers for non–mesurable sets, theorem (5–7), guarantees that these expectation values *can* be realized as frequencies, with very high *probability I.*

The strange character of quantum–statistics will not disappear as a result of such models. The point however is to show that the unintuitive statistical results could be accomodated into an *extension* of probability theory. It is an extention in the formal sense, because all the rules of classical probability theory remain intact in case we deal with Lebesgue measurable sets. We may therefore proceed to construct "realistic" local hidden variable theories for quantum statistics. To see how it is done we shall prove first:

Lemma (5–8): Let L(H) be the lattice of closed projections in the Hilbert space H. There exists a map $E \rightarrow A(E)$ from L(H) to subsets of [0,1] which satisfies:

(a) $A(0) = \phi$

(b) $E < E'$ if and only if $A(E) \subset A(E')$

(c) $A(E \wedge E') = A(E) \cap A(E')$, in particular if $E \wedge E' = 0$

then $A(E) \cap A(E') = \phi$

(d) $A(E) \cup A(E') \subseteq A(E \vee E')$

(e) $\overline{m}(A(E)) = 1$ for all $E \neq 0$

Proof: Let L_1 denote the set of all one dimensional projections in L(H). Then $|L_1| = 2^{\aleph_0}$ and by theorem (5–4) we can associate with each one dimensional projection $L \in L_1$ a sebset $A(L) \subseteq [0,1]$ such that $\overline{m}(A(L)) = 1$, $A(L) \cap A(L') = \phi$

for $L \neq L'$. For the null projection put $A(0) = \phi$ and for $E \in L(H)$, $E \neq 0$ put

$A(E) = \underset{L \leq E}{\cup} A(L)$, then $E < E'$ iff $A(E) \subseteq A(E')$ also $L \leq E \wedge E'$ if and only if

$L \leq E$ and $L \leq E'$ thus $A(E) \cap A(E') = A(E \wedge E')$, also $L \leq E$ or $L \leq E'$

entails $L \leq E \vee E'$ thus $A(E) \cup A(E') \subseteq A(E \vee E')$. Finally since $\overline{m}(A(L)) = 1$

for all $L \in L_1$ we have $\overline{m}(A(E)) = 1$ for all $E \neq 0$.

Note that $A(E)$ and $A(E^{\downarrow})$ are not set theoretical complements though

$A(E^{\downarrow}) \subseteq \widetilde{A}(E)$. Hence no contradiction arises between the above lemma and the

Kochen and Specker theorem. This however will not influence the statistical results

as we shall see shortly. We shall now proceed to define mixtures:

Theorem (5-9): Let $E \to A(E)$ be the association of projections $E \in L(H)$ with

subsets of $[0,1]$ as in Lemma (5-8). Let Σ be the σ-algebra generated by

$\{A(E); E \in L(H)\}$. If W is a state on H there exists a mixture ρ_w on Σ such

that:

(a) $\rho_w(A(E)) = \text{tr}[WE]$ for $E \in L(H)$

(b) $\rho_w(\overset{\infty}{\underset{n=1}{\cup}} A(E_n)) = \text{tr}[W(\overset{\infty}{\underset{n=1}{\vee}} E_n)]$

where $\overset{\infty}{\underset{n=1}{\vee}} E_n$ is the projection on the closed subspace spanned by $\overset{\infty}{\underset{n=1}{\cup}} E_n(H)$.

Proof: For $E \in L(H)$ define $\rho_w(A(E)) = \text{tr}(WE)$ since for all $E \neq 0$ we have

$\bar{m}(A(E)) = 1$, $\underline{m}(A(E)) = 0$ we obtain $0 = \underline{m}(A(E)) \leq \rho_w(A(E)) \leq \bar{m}(A(E)) = 1$.

Also $A(E) \subseteq A(E')$ iff $E < E'$ which entails $\text{tr}(WE) \leq \text{tr}(WE')$ and thus

$\rho_w(A(E)) \leq \rho_w(A(E'))$. We have defined ρ_w on the sets of the form $A(E)$,

$E \in L(H)$.

We shall extend ρ_w to the rest of Σ. Consider first sets of the form

$\bigcup_{n=1}^{\infty} A(E_n)$ where $E_n \in L(H)$. By definition we have $\bigcup_{n=1}^{\infty} A(E_n) \subseteq A(\bigvee_{n=1}^{\infty} E_n)$ where

$\bigvee E_n$ is the projection on the closed subspace spanned by $\cup E_n(H)$. Thus it is

consistent to put $\rho_w(\cup A(E_n)) = \rho_w(A(\bigvee E_n)) = \text{tr}[W(\bigvee E_n)]$. With this definition

we can proceed to extend ρ_w to the rest of Σ by a standard application of Zorn

lemma. Let Σ_0 be the family of all subsets of the form $\cup A(E_i)$, $E_i \in L(H)$ then ρ_w

is defined for Σ_0. Now let $\mathscr{F}$ be the family of all pairs (Γ, ρ) where $\Sigma_0 \subseteq \Gamma \subseteq \Sigma$ and

ρ is defined for every element of Γ and satisfies $\rho(A) \subseteq \rho(B)$ for all $A, B \in \Gamma$ such

that $A \subseteq B$, and also $\underline{m}(A) \leq \rho(A) \leq \bar{m}(A)$ for all $A \in \Gamma$, and the restriction of ρ to

Σ_0 is just ρ_w.

$\mathscr{F}$ is partially ordered by $(\Gamma, \rho) < (\Gamma', \rho')$ iff $\Gamma \subseteq \Gamma'$ and $\rho(A) \leq \rho'(A)$ for all

$A \in \Gamma$. If $\{(\Gamma_i, \rho_i) \; i \in I\}$ is a chain put $\Gamma_I = \cup \; \Gamma_i$, and for $A \in \Gamma_I$ define

$\rho_I(A) = \sup \rho_i(A)$, where the supremeum is taken over all indices i such that $A \in \Gamma_i$.

Clearly $(\Gamma_I, \rho_I) \in \mathscr{F}$ and it is maximal in the chain. By Zorn lemma there exists a

maximal pair (Γ, ρ).

The proof is concluded when we show that $\Gamma = \Sigma$. Indeed if $B \in \Sigma$ but

B ∉ Γ we can extend ρ to Γ ∪ {B} by chosing $\rho(B)$ to satisfy

$\rho(B) \geq \max \{\underline{m}(B), \sup \{\rho(C); C \subseteq B, C \in \Gamma\}\}$

$\rho(B) \leq \min \{\overline{m}(B), \inf\{\rho(D); B \subseteq D, D \in \Gamma\}\}$

in contradiction to the maximality of (Γ, ρ).

Let us examine the hidden variable theory obtained when we associate every quantum property E with the subset A(E) and every quantum state W with a mixture ρ_w.

(a) If E_1, E_2 are two commuting projections then for all W:

$\text{tr}[W(E_1 \vee E_2)] + \text{tr}[W(E_1 \wedge E_2)] = \text{tr}(WE_1) + \text{tr}(WE_2)$ this means that

$\rho_w[A(E_1) \cup A(E_2)] + \rho_w[A(E_1) \cap A(E_2)] = \rho_w(A(E_1)) + \rho_w(A(E_2))$ and therefore by principle III the existence of the properties associated with $A(E_1)$ and $A(E_2)$ can be simultaneously varified on a single sample.

(b)   The hidden variable theory is local.   Consider as an example the Clauser–Horne version of the E.P.R. experiment, Section 3.7.   Let H be the four dimensional space associated with two electrons spin states, E → A(E) the association of subspaces with sets, as above and ρ_s the mixture induced by the singlet state.   We have considered four projections E_1, E_2, E_3, E_4 defined by (3–12), then

$$\rho_s(A(E_i)) = \text{tr}(WE_i) = \frac{1}{2} \qquad i = 1,2,3,4$$

$$\rho_s(A(E_1 \wedge E_3)) = \rho_s(A(E_1 \wedge E_4)) = \rho_s(A(E_2 \wedge E_4)) = \frac{3}{8}$$

$$\rho_s(A(E_2 \wedge E_3)) = 0$$

Thus by principle IV the frequencies observed in the E.P.R. experiment are just these expectation values. The principle of locality is maintained. This is true because $A(E \wedge E') = A(E) \cap A(E')$ for all projections E, E'. Thus if λ is the hidden variable state of the electron pair, then $\lambda \in A(E_1)$ just in case the left electron has spin up in the x direction (independtly of whether a measurement is actually performed) and $\lambda \in A(E_3)$ just in case the right electron has spin up in the z direction (again idenpendently of an actual measurement). Since $A(E_1 \wedge E_3)$ = $A(E_1) \cap A(E_3)$ the E.P.R. experiment will show "spin up in the x direction on the left and spin up in the z direction on the right" precisely in case $\lambda \in A(E_1) \cap A(E_2)$ that is, if and only if the left electron *had* spin up in the x direction before the measurement and the right electron had spin up in the z direction before the measurement. Hence if we had a source of particles, all in the hidden variable state λ, then changing the orientation of the magnet on the right or removing it altogether will not effect the left electron whose properties are completely fixed before the measurement.

This trick can be preformed because of the nature of the concept of probability which is being used. This concept allows for a violation of the classical constraints on frequencies without the violation of locality.

(c) Quantum statistics is accomodated by the hidden variable theory but the lattice operations among observables cannot be accomodated (Kochan and Specker theorem). For example $A(E) \cup A(E') \neq A(E \vee E')$. This however does not influence the statistics since the lattice relations are preserved almost surely, that is in sets whose expectation is one. Consider the Kochen and Specker result, we are dealing with a massive spin-1 particle and its spin space is the three dimensional complex Hilbert space H. Let $E_x \in L(H)$ be the projection on the subspace corresponding to "spin zero in the x-direction" and $A(E_x)$ the corresponding hidden variable subset. If $x_1, x_2, ..., x_{117}$ are the directions in Lemma (4-5) then

$$\bigcap_{\{a_i,a_j,a_k\} \text{ triangle in } \Gamma} [A(E_{x_i}) \cup A(E_{x_j}) \cup A(E_{xk})] \ \cap \ \bigcap_{\{a_i,a_j\} \text{ edge in } \Gamma} [\breve{A}(E_{x_i}) \cup \breve{A}(E_{x_j})] = \phi$$

where Γ is the graph in Fig. (4-2). This is the case because proposition (4-4) is a classical logical falsity. But in our model $A(E) \cup A(E') \subsetneqq A(E \vee E')$. Since $E_{x_i} \vee E_{x_j} \vee E_{x_k} = I$ whenever $\{a_i,a_j,a_k\}$ is a triangle in Γ and since $E^\perp_{x_i} \vee E^\perp_{x_j} = I$ whenever $\{a_i, a_j\}$ is an edge in Γ we have

$$\bigcap_{\{a_i,a_j,a_k\} \text{ triangle in } \Gamma} A(E_{x_i} \vee E_{x_j} \vee E_{x_k}) \ \cap \ \bigcap_{\{a_i,a_j\} \text{ edge in } \Gamma} A(E^\perp_{x_i} \vee E^\perp_{x_j}) = A(I)$$

Since $\rho_w(A(I)) = 1$ the result is compatible with quantum theory.

(d)   Indeed it is clear that our Kolmogorovian models strictly obey classical logic. Given a physical system we shall say that the proposition "the system has property E" is true just in case the hidden variable λ associated with the system is an element of $A(E)$. This certainly is a classical truth value assignment. Since $A(E^\perp) \neq \breve{A}(E)$ we have $A(E) \cup A(E^\perp) \neq [0,1]$ so there are hidden variables λ for which the system has neither the property E nor the property $E^\perp$. But this is not a problem for two reasons: firstly we do *not* identify the property $E^\perp$ with "not E" (in opposition to Birkhoff and von Neumann), secondly the statistic remains intact since $\rho_w(A(E) \cup A(E^\perp)) = \rho_w(A(E \vee E^\perp)) = \rho_w(A(I)) = 1$, so the set of all hidden variables which are either in $A(E)$ or in $A(E^\perp)$ anyway has expectation 1.

The hidden variable theory proposed here lacks any predictive value order and above quantum theory. The motivation for introducing it is not physical but rather logical: to demonstrate that a consistent local hidden variable theory is possible. This is not to say that more elaborate versions of such theories completely lack physical significance. If one introduces dynamics to such theories one can obtain predictions which transcend quantum theory. An example of such possible violations of quantum statistics is provided in Pitowsky (1983). There, a model of spin statistics based on a similar extension of probability is developed, and it is indicated that it is possible, in principle, to manipulate the hidden variables and obtain new results.

5.6 Kolmogorovian Models and Axiomatic Set Theory

The crucial fact which makes our Kolmogorovian models possible, is the existence of nonmeasurable sets. This fact is tightly related to the axiom of choice, as is evident from theorem (5-4). I shall review the status of this axiom first and then analyze its precise relation to nonmeasurable sets.

The standard axiom system for set theory (the axiom of choice excluded) is denoted by Z.F. after Zermello and Fraenkel who first introduced it. From Gödel's famous (second) incompleteness theorem it follows that we cannot prove that Z.F. is consistent, by means available to Z.F. itself, since clearly set theory is more powerful than arithmetic.

Gödel (1940) proved that the axiom of choice (AC for short) is relatively consistent, that is if Z.F. is consistent then Z.F. $+$ AC is consistent as well. Cohen (1966) proved that the axiom of choice is indeed an axiom, and not a theorem of Z.F. In other words, if Z.F. is consistent, then Z.F. $+ \sim$ AC is consistent as well. Formally this means that one can add the axiom of choice to Z.F. without fear that a contradiction will arise, which is not already in Z.F. Alternatively one can replace the axiom of choice by a weaker principle, or by its negation, without fear of a new inconsistancy. How are we to decide which one of these steps to take?

Since no formal criterion can decide whether the axiom of choice is valid or false, we must adopt an informal approach. The most obvious way to go about it is to examine the consequences of the axiom of choice in mathematics.

Suppose first that we drop the axiom altogether, in this case some very strange things may occur. For example there is a model of Z.F. $+ \sim$ AC in which the set of real numbers can be represented as a countable union of countable sets. In such a model the Lebesgue measure is additive but not σ-additive. Similarly there are models of Z.F. $+ \sim$AC in which the sequential and topological defintions of "continuous function" do not coincide. It is therefore obvious that some version of the axiom of choice is indispensable for classical mathematics. The full fledged axiom of choice, however, has some "undesired" and "unintuituve" consequences, in particular the existence of nonmeasurable sets. (The Banach–Tarski paradox is a dramatic case.) The natural question to ask therefore is whether there exists a "middle of the road" approach, which recovers the desired results and rids us of the "paradoxes".

Solovay (1970) constructed a model of Z.F. in which a weaker version of a choice principle applies and in which every subset of the set of reals is Lebesgue measurable (that is, no nonmeasurable sets exist). Solovay's model is also very attractive from a topological standpoint and appears to be that "middle of the road" approach we were looking for.

There is one great problem with Solovay's model, though. In addition to the usual Z.F. axioms Solovay assumed the validity of another esoteric axiom, called "the axiom of inaccessibility" (IN for short), which asserts the existence of an inaccessible cardinal.

The precise formulation and meaning of IN is not important to us. Suffice it to say that IN has a much more dubious status than the axiom of choice itself. The reason is that IN is so powerful that one cannot even prove its *relative* consistency. (This is not due to a lack of ingenuity but rather a matter of principle. Any demonstration that the consistency of Z.F. implies the consistency of Z.F. + IN, will immediately entail that Z.F. is inconsistent.) Again we face a dilemma: Should we adopt the axiom of choice with all its consequences of else adopt Solovay's model with the dubious axiom IN?

Again we may look for a "middle of the road". Maybe we can have the best of all possible worlds. Perhaps we can have a model of Z.F. which is like Solovay's, only that IN is not assumed? This unfortunately is impossible. It turns out that IN is indispensable for any model in which all sets of reals are Lebesgue measurable. This is demonostrated in Shelah (1984).

So we are back to square one. There exists no formal argument against the existence of nonmeasurable sets (since AC is consistent relative to Z.F.) and there exists no informal argument as well. If we want to dispense with nonmeasurable sets we must pay a heavy price, heavier perhaps than the trouble we wanted to dispose of. This means, among other things, that our Kolmogovorian models are not just consistent. There does not even exist an informal mathematical reason for their a–priori rejection.

5.7 Notes and Remarks

A survey of various hidden variable theories can be found in both Belitante (1973) and Bub (1977). The analysis here follows more or less the approach in Kochen and Specker (1967), and is close to the spirit of the contextual hidden variable approach in Gudder (1970) and Shimony (1984).

Theorem 5.4 is taken from Hewitt and Ross (1979). The axiom of choice and its logical status are summarized in Jech (1973). Earlier versions of Kolmogorovian models appear in Pitowsky (1982b, 1983, 1985c, 1986), and their philosophical implications are discussed in Pitowsky (1985a). These models were extended by Gudder (1984a, 1984b, 1985, 1988).

6. Philosophical Remarks

6.1 Physical Realism and Quantum Mechanics

Broadly speaking, physical realism is the doctrine which maintains that some of our physical theories are ture, or approximately ture, and that we can provide a justification for this judgement. All classical physical theories (quantum mechanics excluded, but general relativity included) share some general features. They postulate the existence of certain objects (particles, systems of particles, fields, etc.) the existence of certain properties of these objects (particles having certain mass, electric charge, and so forth) and the existence of certain relations among the objects (distances, relative, velocities, forces). Properties and relations function in causal explanations; thus for example a particle has this or that trajectory because of its mass charge, its initial position and momentum relative to a field. For that reason we shall call these classical theories *causal theories*.

A physical realist, who believes that such a theory is true, is commited to the existence of the objects, properties, and relations postulated by the theory, and to the truthfulness of the causal explanations associated with these properties and relations.

The instrumentalist, in extreme opposition, takes a sceptical view with respect to theories. What matters to the instrumentalist is the directly accessible and directly observable phenomena. Theories are mere instruments of economical organization of experimental facts and for predictions of new facts. The success of some theories, impressive as it may be, is not an indication that the theoretical objects properties and relations postualted by the theory exist, nor that the causal influences assumed by the theory are really there. For example, gravitational fields in Newtonian mechanics are just mathematical objects whose function is to "save the phenomena" that is, to organize, predict and explain (i.e. deduce) such observable facts as the trajectories of celestial bodies, the magnetude of tidal waves, and the like.

The debate over realism has become a fashionable subject in philosophy in the last twenty odd years, so much so that one is almost forced to voice one's educated opinion on the general and broad issues involved. It is not my intention to formulate my own version here. I shall only reflect upon the peculiar role that quantum theory has played in this debate.

Usually agruments over realism are metaphysical and therefore independent of the details and content of scientific theories. Lately, however, some "scientific" arguments against realism have been proposed, arguments which appeal to quantum theory in general and to the violation of Bell's inequality in particular. My aim here is not to argue that physical realism is a viable position, for all I know it may not be. My purpose is just to show that scientific and empirical facts do not support the antirealist (nor the realist for that matter). I think that claiming the opposite comes close to being a category mistake. I find it quite amazing that the antirealist, who believes in the underdetermination of scientific theories by evidence, nevertheless argues that a metaphysical doctrine can be supported by experimental data. You can't eat the cake and have it too.

What we face in fact is a logical and not a metaphysical problem; so rather than addressing "realism" as such, consider a position which I shall call Formal Physical Realism. A formal physical realist accepts a causal theory under the following conditions.

(a) *The theory is, as far as he or she knows, empirically adequate.*

(b) *It is consistent to maintain that all objects, properties and relations which appear in the theory exist, and that the causal influences postulated by the theory obtain in reality.*

The acceptance of a causal theory therefore does not imply that the theory is true or approximately true, only that it is a condidate for being the true depiction of physical reality, in that we can consistently *pretend* that it is true.

The great dream of the antirealist is to find a body of phenomena which cannot be accomodated by any causal formal realist theory. Every such theory will simply be empirically inadequate. If such a body of phenomena exists then realism (in its metaphysical sense) is once and for all dumped in the garbage heap of the history of philosphy, being simply a logically inconsistent position.

Observed microphysical phenomena, predicted by quantum mechanics comes close to realizing the antirealist dream; close, but not quite there. This is in any case the role assigned by the antirealist to quantum mechanics, to demonstrate that realism is not merely bad metaphysics (which perhaps it is), but moreover it is inconsistent or at least incoherent.

The reaction of scientists and philosphers to the strange world uncovered by quantum theory varies dramatically. Some of the more confident and self assured personalities see in it a proof and vindication of their metaphysical beliefs in holism (Bohm 1981), Zen Budhism (Capra 1975), mind–body dualism (Wigner 1961), and what not. I propose to call this mystical approach "the California interpretation of quantum mechanics." Those who are less convinced of the merits of holism and mysticism may tend to be more sceptical. They may say with Feynman: "I do not understand this," by which they probably mean that it seems inconceivable that such phenomena could come about; it seems to defy explanation.

But the one who says "I do not understand" assumes that there is something to be understood, some explanation. Perhaps there is no causal explanation. Maybe phenomena just occur without reason, appear without cause, they are and that is it. This is the core of the antirealist view: we can organize quantum phenomena, we can predict new phenomena, but we cannot provide a causal explanation of it, simply because there is no adequate consistent causal explanation.

In Section 3.10 we have discussed the outline of the antirealist argument. Suppose that we maintain that particles exist and possess properties independently of observation, properties which causally explain their behavior in various experiments. In this framework we are committed to the "balls in the urn" view of statistics and therefore the observed frequencies must obey the classical constraints. Since they often do not, we conclude that quantum "properties" are not causal agents, they are simply names for the phenomena themselves. But this argument is, of course, incomplete. We may explain the violation of the classical constraints by appealing to experimental disturbances, that is by accepting the uncertainty principle in its simplest "operational" form. Consequently, we can adopt a classical causal hidden variable theory, which is formally realist or alternatively a realist quantum logic. In both cases we must reject the princple of locality.

Hence the antirealist argument must be more subtle. The claim is not that *every* causal theory is empirically inadequate, but rather that every local causal theory must fail. Still the argument is quite powerful, for the princple of locality is cherished by the physical realists themselves. It is a central dogma in the very successful and widely accepted special theory of relativity.

But even this version of the argument will not do, for we can resort to causal *local* hidden variable theories which "explain" the phenomena by extending the notion of probability. Such theories were developed in Chapter 5. In these theories the "balls in the urn" view of statistics is maintained and quantum properties are conceived as causes. The theories are certainly consistent and this is all which is required by "formal realism."

In short, nothing metaphysical is implied by quantum phenomena. The experimental results are consistent with every conceivable view. In this respect, quantum mechanics is just like other theories, past and present, and there is no rational reason why it should play a peculiar role in debates over realism. As much as we can consistently pretend that gravitational fields exist we can pretend that "spin value in the x–direction" exists (independently of observation). Philosophically , it is the same kind of pretence.

The debate over realism will not be added by introducing "scientific" considerations and certainly not by experimental data. In fact, pseudo- scientific arguments tend to obscure the true nature of the debate. Hume has made the sceptical point regarding causes in as forceful a manner as it can be made. Quantum theory has added nothing new to it.

But are hidden variable theories *physically* significant? Can we produce a hidden variable theory which goes beyond quantum theory and predicts new phenomena? Psychologically this question seems to be connected to the problem of realism. It seems that realists are more inclined to consider hidden variable theories. Logically however, the physical merits of hidden variable theories are independent of the metaphysical debate. One can take an instrumentalist position regarding hidden variables and still appreciate their potential application in predicting new phenomena. Alternatively one can believe in the existence of hidden causes, while still holding on to the view that these causes lie beyond our powers of manipulation.

Hidden variable theories may very well lead to new predictions, this is a matter of empircal judgment. My own bias tends to side with non-local theories such as Bohm (1952). In this and similarly detailed dynamical theories we may be able to find a way to manipulate and control the hidden variables and arrive at a new phenomenological level. In any case, I can conceive of no a-priori argument why this could not occur.

6.2 Quantum Theory and the Foundations of Probability

The proper interpretation of the term "probability" has been a subject of philosophical controversy since the seventeenth century. The central approaches which emerged in our century are the frequencist school (Von Mises, 1957; Popper, 1959), the logical approach (Keynes, 1943; Carnap, 1950), and the subjectivist school (Ramsey, 1926; De Finetti, 1972). All these schools accept the validity of the axioms of probability theory, they differ with respect to the interpretation and the justification of these axioms.

Regardless of one's view on the meaning and justification of the axioms of probability, it is obvious that these axioms entail that the values of probabilities should be subjected to the constraints which were developed in Chapter 2. In other words, probabilities should satisfy the facet inequalities of c(n,S). Indeed, we have seen that these inequalities have different meaning for the different approaches. In the frequency approach the inequalities represent a–priori constraints on proportions of properties. For the logical and subjectivist camp the inequalities represent a–forteriori constraints on averages of truth values. Be it as it may, the violation of the inequalities by quantum frequencies poses a problem to all schools.

With almost no exception this problem has been ignored by probablists. There are various reasons for this, the most important one has to do with the technical character of the results. Bell type inequalities were derived by physicists in the context of an analysis of certain concrete experiments. This has created the impression that the constraints in question have something to do with the physical world, or with quantum mechnaics in particular. Consequently, the violation of the inequalities appeared as an esoteric physical phenomenon. Once one realizes that the character of the constraints is directly related to the traditional concept of probability, and has nothing to do with physics as such, the dimensions of the dilemma, and its relations to the foundations of probability become evident.

There is one obvious way to go about and solve the problem. One can argue that there is nothing wrong with our traditional concept of probability. The apparent violation of its rules is caused by "measurement distrubances." Now, quantum theory does not speak aobut measurment disturbances but rather about "interference," and the meaning of this term is precisely what is at stake. Quantum mechanics does not provide any dynamic *mechanism* of measurement disturbances. If one introduces such a mechanism one goes a step beyond quantum theory and obtains a non local theory (theorem (5-3)). As yet there is no independent evidence for the existence of such a mechanism, the only evidence is the violation of Bell inequalities themselves. Thus the dilemma is to chose between non local theories and the axioms of probability. Presently no rational guideline for such a choice seems to be available.

In any case, independently of quantum theory, one can conceive of consistent "possible worlds" in which the axioms of probability fail to obtain with respect to observed frequencies. The Kolmogorovian models of Sections 5.4 and 5.5 can serve as examples. Hence it is at least clear that the usual axioms of probability theory cannot be taken as "analytic." In other words, the axioms do not necessarily capture the proper meaning of the term "probability" (though they surely represent a possible analysis of the term).

This conclusion is most devastating for the logical camp, Carnap's approach in particular. Carnap (1950) took the axioms of probability as mere conclusions of propositional logic. As such, these axioms should be valid in all possible worlds. Since they are not, then either one has to give up the logical conception of probability, or else to deny the analytic character of logic. Both alternatives undermine Carnap's position.

The subjectivist is also hard hit by our conclusion. The subjectivist typically resorts to a "Dutch book" argument in order to justify the a–priori character of the axioms of probability. These arguments have the following structure: Suppose that an agent is consistently violating one of the axioms of probability; then we can design a gamble ("sell the agent a Dutch book"), such that the agent is bound to lose his bet. Now consider the following question: Is it possible to construct a gambling device (for example, roulette wheel of some kind), such that an agent is bound to lose, in the long run, if s/he is betting *in accordance* with the principles of classical probability?

Surely one cannot rule out the existence of such a device on the basis of any analytic or a purely logical argument. We know that there are possible worlds in which such a device is possible (again see the Kolmogorovian models). The subjectivist may argue that the laws of physics in the actual world prevent us from constructing such a "roulette wheel." In this case, the subjectivist admits that probability is a synthetic concept. The irony is that quantum theory indicates that strange gambling devices *may* exist in this very actual world.

The argument is very simple. Take our gambling device to be an E.P.R. experiment as in the Clauser–Horne version of Section 3.7. The agent is informed about the set–up of the experiment, that is about the orientation of the magnets (for example, x–direction on the left, z–direction on the right). Then the agent is faced with the outcome on the left hand side ("up" or "down"), and has to bet on the outcome on the right hand side.

If the reader is dissatisfied with the overly technical appearence of the device we can disguise the entire set up, make it simpler and more commerically appealing. For example, the outcomes can be represented by the flashing of colored bulbs, and the magnet orientations can be represented by the position of bright dials.

We assume that the agent who participates in the game is ignorant of quantum theory but is nevertheless rational (there are such people!). Now suppose that God appears to the agent and promises him that no signal or connection of any kind passes between the two sides. Being a rational person, our agent distributes coherent prior probabilities over all the eight possible outocmes of the four possible set–ups. As the game progresses, for some instances, the agent modifies his estimates according to Bayes rule. This does not seem to help, and the agent keeps loosing money at a rate much higher than expected even on a bad day. The reason is that the agent had (rationally) assumed that all eight events could be represented on a usual probability space, and thus his estimations automatically satisfy the Clauser-Horne inequality. Unfortuantely it is violated.

Having had his pockets emptied our agaent may conclude that he was decieved by God. Apart from being an outright blashphamy, this conclusion is unwarranted because there exists another logical possiblity, namely that the axioms of probability do not apply in this case. As long as the only evidence for "non-local disturbances" is the violation of the classical constraints themselves, we cannot safely say that such disturbances do exist and that our agent was really deceived. There is always a possibility that the E.P.R. correlations exist for no cause whatever (the antirealist view) or because of some strange nonmeasurable distribution of properties (the Kolmogorovian models). At any rate, the subjectivist faces a real dilemma.

The frequencist takes the axioms of probability to be empirical truths about relative frequencies of events in the world. This approach suffers from considerable logical setbacks, because of its use of infinite samples, and because of its notion of "randomness" (see e.g. Bub and Pitowsky, 1985). These difficulties notwithstanding, the frequencist can adjust his/her views to accomodate new experimental evidence. If the laws of probability are empirical, not a-priori, they can be modified. The frequencist approach is thus less hard hit by quantum theory or the Kolmogorovian models.

But there is another interpretation of probability which, given the above analysis, should be taken more seriously. In his Encyclopedia of Philosophy article on probability (1967), Max Black provides the following short dismissal of the view which he calls "mathematical dogmatism":

"Brief mention should be made of a view of a kind sometimes found in mathematical textbooks on probability and often recommended in the classroom. Roughly speaking we are asked to conceive of probability as whatever can satisfy the axioms of the mathematical theory. The pure theory of chances is compared to pure geometry, both viewed as "idealized models" having only loose connection with reality, and the task of correlating the precise mathematical results with their imprecise counterparts in experience is held to be basically a practical one, needing no theoretical discussion. As a solution to the problem of interpretation [of probability] this approach is merely an evasion. Formal obeisance to the "idealization" implict in mathematical theory construction serves merely as an excuse for shirking the hard work of articulating the links between theory and practical applications."

This view, which should be called formalism, rather than mathematical dogmatism, is in fact a very serious view. There are "precise mathematical results" which indicate that frequencies may very well disobey the classical constraints. Given this fact no amount of hard work will suffice to articulate the links between the classical axioms and some practical applications. I frankely do not see any difference of principle between this case and pure geometry, some of whose models are Euclidean and some non- Euclidean. What kind of an a-priori argument can justify the application of any? I do agree that the relations between pure mathematical theories on the one hand, and physical reality on the other, require philosophical analysis. But probability has no special status in this respect, over and above group theory, for example.

The trap into which Black, as well as others, fall is old. They believe that philosophical analysis and perhaps metaphysical deliberations, will depict some "adequate" interpretation of a concept, and an a-priori justification for its use in scientific contexts (and maybe outside of science as well). But there is no synthetic a-priori truth. I do not know how many examples are required to drive this lesson home, probability is just one example in a multitude.

Personally, I have always been puzzled by the preoccupation of philsophers with the notion of probablity. Indeed nobody attempts to "justify" the axioms of group theory, Lee algebras or the theory of Hilbert spaces. Surely, there is a good historical reason for this difference in attitude. Probability is afterall the basis of inductive and statistical inference and of rational decision making. One would not want one's inferences and decisions to be model relative. But this, I think, is an irrational attitude. Our expectations do depend on scientific laws, so why shouldn't our inferences and decisions manifest the same dependence? If I am to bet on the outcome of an E.P.R. experiment I will naturally consult quantum theory. If the result is in appearent conflict with the classical patterns of statistical inference, then so much the worse for the classical inference. One ought to follow the rule which works and not the rule which ought to have worked.* I believe that the strength of the natural sciences derives from their adoption of this dictum, regardless of justification and prior to it.

6.3 Mathematical Models and Physics

The application of mathematical rules and theories in physics requires no justification, but it does call for analysis. Indeed the relations between models and reality have lately become a source of a new form of scepticism.

Traditional scepticism typically appeals to the poverty of human perception and limited intellectual resources. We cannot rely on our senses for they often mislead us. We cannot logically justify our expectations, for past regularity does not logically entail its future persistence. Only God, with (potential or actual) infinite capabilities, can percieve the laws of Nature through the appearnces. Modern forms of sceptisism, by contrast, appeal to the unlimited and unbounded human imagination. Our free minds are so flexible that they can invent a multitude, potentially infinite number of theories, which account equally well for the same set of facts. Since these models are pairwise incompatible, and since they all account for everything perceivable, there is no way to decide which model is the true model. Therefore the very claim that there is a "true theory" is incoherent. Moreover, it is not clear what "a fact" means any more. All we have are "facts

---

*This is also my position with respect to Newcomb's paradox. The similarity between this paradox and quantum mechanics is not an accident.

under description" and the description is model relative. This is the source of the thesis known as "The Underdetermination of Theory by Observation" (Quine, 1975). A version of this thesis, applied directly to the problem of realism, can be found in Putnam (1983).

I have already indicated that there is some tension between these views and the older forms of scepticism. Perhaps it is true that we cannot logically justify our expectations on the basis of past experience. Perhaps it is also true that causes exist only in the eye of the beholder. But we can always consistently *pretend* that causes exist, and that past experience counts as solid ground for the future, for we can always invent an empirically adequate theory which implies just that. Scepticism is what it is, doubt and uncertainty; and doubt is a double edged sword which can be applied to itself.

Modern scepticism is "modern" because it could not have been developed before the twentieth century. The realization that theories could not be determined by observation is a result of developments in logic, mathematics and physics in the late nineteenth and early twentieth centuries. It is historically rooted in the development of non–Euclidean geometries and in the revolution in the foundations of mathematical analysis (the calculus).

Non–Eulidean geometry provided the first example of a mathematical construction, which is perfectly consistent, and which defies pre–analytic intuitions with respect to space. Such constructions could have been dismissed as mere abstract games, but they were not: "I am coming more and more to the conviction," wrote Gauss (1817) "that the necessity of our geometry cannot be demonstrated, at least neither by nor for, the human intellect ... geometry should be ranked, not with arithmetic which is purely aprioristic, but rather with mechanics." Legend has it that Gauss conducted experiments with light rays, trying to decide whether space is Euclidean (flat) of else has some small curvature. He allegedly concluded that this matter cannot be decided on the basis of his data (E.T. Bell, 1973).

The work on the foundations of the calculus by Cauchy, Bolzano and in particular Wierstrass has had a different effect. Wierstrass was motivated by the Greek ideal of pure mathematics which requires no physical intution for its justification. His explicit motivation was to free the calculus from physical intuitions which were evident in the Newtonian concept of "flux". Indeed, some of the basic distinctions, nowadays common in analysis, were obscured before his time. "Continuity," "differentiability," and "smoothness" were used without ever being rigorously defined, a practice which gave rise to some amusing controversies. Wierstrass's famous $\varepsilon-\delta$ method put an end to this.

Once freed from its physical origins (or alleged origins), "pure" calculus could deal with such "unphysical" or "unituitive" notions as "continuous but nowhere differentiable curves." These were supposed to be the mark of pure mathematical thinking, creations of the mind, distinguishable from any physical trajectory.

History however is full of ironies (which some call "dialectical processes"). Those concepts which were introduced as symbols of pure mathematical thinking were very often appropriated by later generations as instruments of physics.

Non–Euclidean geomtry is a well known paradigm case. "Continuous nowhere differentiable functions" is a somewhat less known case, and the story is instructive:

After the discovery of the Brownian motion, physicists quickly realized that this phenomenon ought to be explained in terms of the collisions between the Brownian particle, and the molecules of the medium in which it is suspended. Gouy attempted to measure the average velocity of the Brownian particle and derive Avogadro's number. His results were off mark by five orders of magnitude. Einstein obtained a correct prediction in the famous 1905 paper on the Brownian motion. In his analysis he bypassed the reference to the particle velocity. Why was Einstein successful and Gouy so mistaken? Wiener provided the explanation. His mathematical theory of the Brownian motion is one of the origins of the theory of stochastic processes, and modern probability theory in general. Einstein's argument, noted Wiener, is in fact a probabistic one. The probability space in question is the set of all possible continuous trajectories of the Brownian particle, and the basic events are the presence of the particle in a given region in space at a given time. On the space generated by these events Wiener defined a probability measure, the Wiener measure. Then he proved (together with Paley): "*With probability one, the path taken by a Brownian particle is nowhere defferentiable.*"

Now we understand why Gouy failed. The Brownian particle simply does not have a velocity (which is, by definition, the derivative of the path). Needless to say, the Einstein–Wiener model of the Brownian motion is an idealization, which breaks down when we consider segments of the path which are of order of magnitude of molecular dimensions. Both authors were explicitly aware of that.

The concept of "continuous, nowhere differentiable curves" is closely connected to the theory of Fracktals (Mandelbro t, 1977). Judging from the *Physical Review*, it is currently widely used for the mathematical modeling of chaotic phenomena.

Non-Euclidean geometries and nowhere differentiable continous curves are just two examples of pure mathematical constructions which were developed for their own sake and were lately appropriated by physicists. The theory of groups, the theory of invariants, differential topology are other known examples. The distinction between "pure" and "applied" mathematics is forever a temporal one. Today's "purest" theory may be applied tommorow.

Apart from introducing uncertainty with respect to the boundaries of "legal adequate" physical modeling, the development of modern mathematics and logic has had a far more dramatic effect on the level of methodology. The development of modern logic was tightly related to advances in geometry. The first rigiourous formalization of a theory is Hilbert's "Gudlaggen der Geometrie". Cantor's set theory originated in problems of Fourier analysis. The nineteenth century has freed the genie from its bottle.

Two results in mathematical logic are relevant in the present discussion. The first is the Skolem Lowenheim theorem. It states that a first order theory, which has an infinite model, has an infinite model in every (infinite) cardinality. This means that a (first order) formulation of a theory cannot fix an interpretation, at least as far as the cardinality of the "intended model" is concerned. The second, and far deeper result, is Gödel's incompleteness theorem. A sufficiently rich consistent theory, whose set of premises (axioms) is recursive, is never complete.

This means that no recursive set of assumptions can determine an "intended interpretation" even in a fixed cardiality. As is well known, this statement is already valid for number theory.

The Underdetermiantion of Theory by Observation is a methodological consequence of these formal results. Any reasonable set of "laws of Nature," in conjunction with any finite (or even recursive) set of "observation statements" has many incompatible interpretations (provided it is consistent). These interpretations agree on all observable facts and all the "Laws of Nature" are valid in them; though they probably *mean* different things in different models. Yet there are some statments which are valid in one such model and false in another interpretation. The question "which of these models represents the actual world?" has no answer in principle and is therefore perhaps meaningless.

The Kolmogorovian models of chapter 5 exemplify this state of affaris. We can have models of quantum statistics in which all observable statistical results are kept intact. Still the model provides a causal local "realistic" explanation of them. The existence of such models should come as no surprise, given the general theorems cited above.

It seems at first glance that all gates are opened and all we are left with is our free will. Empirical adequacy turns out to be a very weak criterion indeed, it does not provide any deep meaning to our statments. We can, if we like, see in the world an eternal mechanical harmony of causes and effects. Alternatively we can observe the universe through the spectacles of chaos and uncertainty. Nothing in physics seems to indicate the true alternative. Nothing seems to provide an answer to those deep and important questions which motivate research in the first place.

Yet, over and above our free will, we are constrained by logic and by Nature. It seems superfically that logic too is not immune to conceptual change, and quantum logic perhaps is a case in hand. But as we have seen, quantum logic, in its serious realistic conception, is nothing but a hidden variable theory in disguise (Sections 4.4, 4.5).

The idea that logic may change as a result of developments in the empirical sciences is allegedly suggested in the work of Quine (1953). In his conception any of our beliefs can be altered to accomodate new "sense experience." But there is an obvious criterion attached to that claim, namely that the new system of beliefs be consistent (at least not manifestly inconsistent). If there is any meaning to the idea of "changing logic," without giving up bivalence, it is the admission that a classical logical falsity may sometimes be true (or equivalently, a classical tautology may sometimes be false). Hence a system of belief, obtained by changing classical logic while maintaining bivalence, cannot be *classically* consistent. How are we to make sense of the claim that our new belief system is nevertheless consistent? One might answer: "it should be consistent according to the rules of the new logic." But then there is no criterion at all. We can simply declare every set of statements to be consistent according to the rules of some alleged "new logic."

The only change in classical logic which does make sense is giving up bivalence. This has been done by the intuitionists and for reasons that have nothing to do with the empirical sciences. No "sense experience" can be of any help in deciding the battle between intuitionism and classical mathematics. Any other "change in logic" comes close to a misunderstanding, as much as denying that $2 + 2 = 4$ is a misunderstanding (Wittgenstein, 1969).

Apart from logic, science is also constrained by Nature; our theories must be empirically adequate. Logic and Nature are the only constraints. One might attempt to add further criteria, such as simplicity, but these are notoriously difficult to formulate precisely. Ideas which are common practice today would have seemed utterly unintuitive and surely not "simple" in the eyes of our predessessors. The idea that "continous nowhere differentiable curves" would some day be instrumental in physics would have seemed absurd to Wierstrass and his peers. By analogy, we tend to think that the continuum hypothesis will never be used instrumentally in theoretical physics. But is there any a–priori argument to the effect that this could never occur?

Still, the aim of science is to search for the truth. I believe that this normative statement implies no ontological commitment to the existence of The True Ideal Theory. All science seeks is that magic point, where Nature, and our perceptual and intellectual resources, finally reach their equilibrium.

References

Aspect, A., Grangier, P. and Roger, G. (1981), Experimental tests of
   realistic local theories via Bell's theorem, *Phys. Rev. Lett. 47*,
   460, reprinted in Wheeler and Zurek (1983).

Belifante, F.J. (1973), *A Survey of Hidden-Variable Theories*,
   Pergamon, Oxford.

Bell, E.T. (1973), *Men of Mathematics*, Simon and Schuster, New York.

Bell, J.S. (1964), On the Einstein-Podolsky-Rosen paradox, *Physics 1*,
   195.

Beltrametti, E.G. and van Fraassen, B.C. (1981) eds., *Current Issues in
   Quantum Logic*, Plenum, New York.

Birkhoff, G. and von Neumann, J. (1936), The logic of quantum mechanics,
   Ann. Math. 37, 823.

Black, M. (1967), Probability, *Encyclopedia of Philosophy*, Macmillan,
   New York.

Bohm, D. (1951), *Quantum Theory*, Prentic–Hall, Englewood Cliffs, NJ.

Bohm, D. (1952), A suggested interpretation of quantum theory in terms of "hidden variables", *Phys. Rev. 85*, 180, reprinted in Wheeler and Zurek (1983).

Bohm, D. (1981), *Wholeness and the Implicate Order*, Rutledge and Kegan, London.

Bohr, N. (1949), Discussions with Einstein on epistemological problems in atomic physics, in P.A. Schlipp ed., *Albert Einstein Philosopher–Scientist*, The Library of Living Philosphers, Evanston, reprinted in Wheeler and Zurek (1983).

Bonferroni, C.E. (1936a), Teorie statistica delle classi e calcolo delle probabilita, *Public Inst. Sup. Sc. Ec. e Comm. di Firenze 8*, 1.

Bonferroni, C.E. (1936b), Il calcolo delle assicurazioni su grappi di teste, *Studi in Onore del Prof. S.O. Carboni*, Roma.

Bub, J. (1977), *The Interpretation of Quantum Mechanics*, Reidel, Dordrecht

Bub, J. and Pitowsky, I. (1985), Critical notice on K.R. Popper's
    postscript to the logic of scientific discovery, *Can. J. of Phil.* *15*, 539.

Capra, F. (1975), *The Tau of Physics*, Shambahala, Berkeley, CA.

Carnap, R. (1950), *Logical Foundations of Probability*, U. of Chicago
    Press, Chicago, IL.

Clauser, J.F. and Horne, M.A. (1974), Experimental consequences of objective
    local theories, *Phys. Rev. D. 10*, 526.

Clauser, J.F. and Shimony, A. (1978), Bell's theorem. Experimental tests and
    implications, *Rep. Prog. Phys. 41*, 1881.

Clauser, J.F., Horne, M.A., Shimony, A. and Holt, R.A. (1969), Proposed
    experiment to test local hidden–variable theories, *Phys. Rev. Lett.*
    23, 880, reprinted in Wheeler and Zurek (1983).

Cohen, P. (1966), *Set Theory and the Continuum Hypothesis*, Benjamin,
    New York.

Cooke, R., Keane, M. and Moran, W. (1985), An elementary proof of Gleason's
    theorem, *Math. Proc. Comb. Phil. Soc. 98*, 117.

d'Espangant, B. (1971), *Conceptual Foundations of Quantum Mechanics*, W.A. Benjamin, Menlo Park, CA.

De Finnetti, B. (1972), *Proabability Induction and Statistics*, Wiley.

Dummett, M. (1978), Is logic empirical?, in *Truth and Other Enigmas*, Harvard University Press, Cambridge, MA.

Einstein, A., Podolsky, B. and Rosen, W. (1935), Can quantum mechanical description of physical reality be considered complete?, *Phys. Rev.* *47* 777, reprinted in Wheeler and Zurek (1983).

Fine, A. (1982a), Hidden variables, joint probability and Bell inequalities, *Phys. Rev. Lett.* *48*, 291.

Fine, A. (1982b), Reply to Garg A. and Mermin N.D., *Phys. Rev. Lett.* *49*, 243.

Finkelstein, D. (1962), The logic of quantum physics, *Trans. New York Acad. of Sci.* *25*, 621.

Finkelstein, D. (1968), Matter, space and logic, in R. Cohen and M. Wartofsky, eds., *Boston Studies in the Philosophy of Science 5*, 199, Reidel, Dordrecht.

Fréchet, M. (1940), *Les Probabilités Associées a un Système D'Événtments Compatibles et Dépandants*, Hermann, Paris.

Garey, M.R. and Johnson, D.S. (1979), *Computers and Intractability, A Guide to the Theory of NP—Completeness*, W.H. Freeman, New York.

Garg, A. and Mermin, N.D. (1982a), Bell inequalities with range of violation that does not diminish as the spin becomes arbitrarily large, *Phys. Rev. Lett. 49*, 901.

Garg, A. and Mermin, N.D. (1982b), Comment on "Hidden variables, joint probability and Bell inequalities", *Phys. Rev. Lett. 49*, 242.

Garg, A. and Mermin, N.D. (1983), Local realism and measured correlations in the spin s Einstein—Podolsky—Rosen experiment, *Phys. Rev. D. 27*, 339.

Garg, A. and Mermin, N.D. (1984), Farkas's lemma and the nature of reality:
Statistical implications of quntum correlations, *Found. of Phys.*
14, 1.

Gauss, K.F. (1831), Letters to Olbers, *Collected Works vol. II*, 177.

Gleason, A.M. (1957), Measures on the closed subspaces of a Hilbert space,
J. Math. and Mech. 6, 885.

Gödel, K. (1940), The consistency of the axiom of choice and the generalized
continuum hypothesis, *Annals of Math 3*, Princeton U. Press,
Princeton, NJ.

Gudder, S.P. (1970), On hidden variable theories, *J. Math. Phys. 11*,
481.

Gudder, S.P. (1984a), Probability manifolds, *J. Math. Phys. 25*, 2397.

Gudder, S.P. (1984b), Reality, locality and probability, *Found. of Phys.*
14, 997.

Gudder, S.P. (1985), Amplitude phase space model for quantum mechanics,
Int. J. of Theor. Phys. 24, 343.

Gudder, S.P. (1988), *Quantum Probability*, Academic Press, Orlando, FL.

Hewitt, E. and Ross, K.A. (1979), *Abstract Harmonic Analysis*,
Springer, Berlin.

Hume, D. (1739), *A Treatise of Human Nature*, modern edition: D.G.L.
Macnabb ed., Fontana/Collins, Glasgow.

Jech, T.J. (1973), *The Axiom of Choice*, North Holland, Amsterdam.

Jauch, J.M. (1968), *Foundations of Quantum Mechanics*, Addison–Wesley,
Reading, MA.

Karp, R.M. and Papadimitriou, C.H. (1980), On linear characterizations of
combinatiorial optimization problems, *Proc. of the 21 Symp. Found.
Comput. Sci.*, 1.

Keynes, J.M. (1943), *A Treatise on Probability*, McMillan, London
(originally published in 1923).

Kochen, S. and Specker, E.P. (1967), The problem of hidden variables in
quantum mechanics, *J. Math. and Mech. 17*, 59.

Macdonald, A.L. (1982), Comment on "Resolution of the Einstein–Podolsky–Rosen and Bell paradoxes", *Phys. Rev. Lett. 49*, 1214.

Mandelbrot, B. (1977), *Fractals – Form, Chance and Dimension*, W.H. Freeman, San Francisco.

Mermin, N.D. (1982), Comment on "Resoultion of the Einstein–Podolsky–Rosen and Bell paradoxes", *Phys. Rev. Lett. 49*, 1215.

Mermin, N.D. and Schwarz, G. (1982), Joint distributions and local realism in the higher spin Einstein–Podolsky–Rosen experiment, *Found. of Phys. 12*, 101.

Messiah, A. (1963), *Quantum Mechanics*, North Holland, Amsterdam.

Mott, N.F. and Massey, H.S.W. (1965), *The Theory of Atomic Collisions*, Claredon Press, Oxford.

Pitowsky, I. (1982a), Substitution and truth in quantum logic, *Philos. of Sci. 49*, 380.

Pitowsky, I. (1982b), Resolution of the Einstein–Podolsky–Rosen and Bell paradoxes, *Phys. Rev. Lett. 48*, 1299.

Pitowsky, I. (1983), Deterministic model of spin and statistics, *Phys. Rev. D. 27*, 2316.

Pitowsky, I. (1985a), On the status of statistical inferences, *Synthese 63*, 233.

Pitowsky, I. (1985b), Discussion: Quantum mechanics and value definiteness, *Philos. of Sci. 52*, 154.

Pitowsky, I. (1985c), A phase space model of quantum mechanics in which all operators commute, in L.M. Roth and A. Inomata, eds., *Foundamental Questions in Quantum Mechanics*, Gordon and Breach, New York.

Pitowsky, I. (1986), The range of quantum probability, *J. Math of Phys. 27*, 1556.

Pitowsky, I. (1988), Correlation polytopes, their geometry and complexity, forthcoming.

Popper, K.R. (1959), *The Logic of Scientific Discovery*, Basic Books, New York (German original 1934).

Putnam, H. (1968), Is logic empirical? in R. Cohen and M. Wartofsky, eds., *Boston Studies in the Philosophy of Science, Vol. 5*, Reidel, Dordrecht.

Putnam, H. (1976), How to think quantum logically, in P. Suppes, ed., *Logic and Probability in Quantum Mechanics*, Reidel, Dordrecht.

Putnam, H. (1983), Models and reality, in *Realism and Reason*, Cambridge U. Press, Cambridge.

Quine, W.V.O. (1953), Two dogmas of empiricism, in *From a Logical Point of View*, Harper and Row, New York.

Quine, W.V.O. (1975), On empirically equivalent systems of the world, *Erkenntnis IX*, 313.

Ramsey, F.P. (1926), Truth and probability, in R.B. Brathwaite, ed., *The Foundations of Mathematics*, Routledge and Kegan, London.

Reichenbach, H. (1944), *Philosophical Foundations of Quantum Mechanics*, University of California Press, Berkeley.

Rockafeller, R.T. (1970), *Convex Analysis*, Princeton University Press, Princeton.

Shaeffer, T.J. (1978), The complexity of satistiability problem, *Proc. 10th. Ann. Syp. on Theory of Computing*, 216, Association for Computing Machinary, New York.

Shelah, S. (1984), Can you take Solovay's inaccessible away? *Israel J. of Math.* *48*, 1.

Shimony, A. (1984), Contextual hidden variable theories and Bell inequalities, *Brit. J. Phil. Sci.* *35*, 25.

Solovay, R.M. (1970), A model of set theory in which every set of reals is Lebesgue measurable, *Ann. of Math.* *92*, 1.

Stairs, A. (1983), Quantum logic, realism and value definiteness, *Phil. of Sc.* *50*, 578.

van Fraassen, B.C. (1980), *The Scientific Image*, Oxford U. Press, oxford.

van Fraassen, B.C. (1982), The Charidbis or realism: Epistemological implications of Bell's inequality, *Synthese 52*, 885.

Varadarajan, V. (1962), *Geometry of Quantum Theory, Vol. I, II*, van Nostrand, Princeton, NJ.

von Mises, R. (1957), *Probability Statistics and Truth*, Dover, New York (original German edition, 1928).

von Neumann, J. (1955), *Mathematical Foundations of Quantum Mechanics*, Princeton University Press, Princeton (original German edition, 1933).

Wheeler, J.A. and Zurek, W.H. (1983), eds., *Quantum Theory and Measurement*, Princeton University Press, Princeton.

Wigner, E.P. (1961), Remarks on the mind–body question in *Symmetrics and Reflections*, Indiana University Press, Bloomington, reprinted in Wheeler and Zurek, 1983.

Wigner, E.P. (1970), On hidden variables and quantum mechanical
    probabilities, *Amer. Jour. Phys.* *38*, 1005.

Wigner, E.P. (1976), Interpretation of quantum mechanics, mimeographed note,
    reprinted in Wheeler and Zurek, 1983.

Wittgenstein, L. (1939), *On Certainty*, eds.: E.M. Enscomb and G.H. von
    Wright, Blackwell, Oxford.

Yemlichev, A.V., Kovalev, M.M. and Kravtsov, M.K. (1984), *Polytopes,
    Graphs and Optimizations*, Cambridge University Press, Cambridge
    (Russian original, 1981).

K. V. Laurikainen, University of Helsinki, Finland

Beyond the Atom

The Philosophical Thought of Wolfgang Pauli

1988. XIX, 234 pages. ISBN 3-540-19456-8

Contents: Introduction. – Positivism and Realism. – The Reality of
Opposites. – The Metaphysical Roots of Science. – Spirit and Matter.
– The Limits of Knowledge. – Mysticism. – The Problem of Evil. –
Quaternity. – Transcendental Reality. – Appendices: – Wolfgang Pauli
and the Copenhagen Philosophy. The Role of the Observer in Micro-
physics. The Possibility of Science and Its Limits. Translations of the
German Quotations into English. – Subject Index.

G. Ludwig, University of Marburg, Germany

An Axiomatic Basis for Quantum Mechanics

Volume 1
Derivation of Hilbert Space Structure

1985. 6 figures. X, 243 pages. ISBN 3-540-13773-4

Contents: The Problem of Formulating an Axiomatics for Quantum
Mechanics. – Pretheories for Quantum Mechanics. – Base Sets and
Fundamental Structure Terms for a Theory of Microsystems. –
Embedding of Ensembles and Effect Sets in Topological Vector
Spaces. – Observables and Preparators. – Main Laws of Preparation
and Registration. – Decision Observables and the Center. – Represen-
tation $\underline{B}$, $\underline{B'}$ by Banach Spaces of Operators in a Hilbert Space. –
Appendices. – Bibliography. – List of Frequently Used Symbols. – List
of Axioms. – Index.

Volume 2
Quantum Mechanics and Macrosystems

1987. 4 figures. IX, 242 pages. ISBN 3-540-17540-7

Contents: Further Structure of Preparation and Registration. – The
Embedding Problem. – Compatibility of PT_q with PT_qexp. – Special
Structures in Preparation and Registration Devices. – Relations
between Different Forms of Quantum Mechanics and the Reality
Problem. – Bibliography. – List of Frequently Used Symbols (1). –
List of Frequently Used Symbols (2). – List of Axioms. – Index.

In preparation
W. Schommers, Karlsruhe (Ed.)

Quantum Theory and Pictures of Reality

Foundation, Interpretations, and New Aspects

With contributions by B. d'Espagnat, P. Eberhard, W. Schommers,
F. Selleri

1989. 31 figures. Approx. 300 pages. ISBN 3-540-50152-5

Springer-Verlag Berlin
Heidelberg New York London
Paris London Hong Kong

Lecture Notes in Mathematics

Lecture Notes in Physics